Modeling Love Dynamics

WORLD SCIENTIFIC SERIES ON NONLINEAR SCIENCE

Editor: Leon O. Chua
University of California, Berkeley

---

*To view the complete list of the published volumes in the series, please visit:
http://www.worldscientific.com/series/wssnsa

WORLD SCIENTIFIC SERIES ON
NONLINEAR SCIENCE

Series A   Vol. 89

Series Editor: Leon O. Chua

Modeling Love Dynamics

Sergio Rinaldi

Politecnico di Milano, Italy
International Institute for Applied Systems Analysis, Austria

Fabio Della Rossa
Fabio Dercole
Alessandra Gragnani

Politecnico di Milano, Italy

Pietro Landi

Stellenbosch University, South Africa

World Scientific

NEW JERSEY · LONDON · SINGAPORE · BEIJING · SHANGHAI · HONG KONG · TAIPEI · CHENNAI · TOKYO

Published by

World Scientific Publishing Co. Pte. Ltd.

5 Toh Tuck Link, Singapore 596224

USA office: 27 Warren Street, Suite 401-402, Hackensack, NJ 07601

UK office: 57 Shelton Street, Covent Garden, London WC2H 9HE

British Library Cataloguing-in-Publication Data
A catalogue record for this book is available from the British Library.

Cover Credit: Part of "The Kiss" by Gustav Klimt (1908)
Courtesy of Österreichische Galerie Belvedere, Vienna.

World Scientific Series on Nonlinear Science Series A — Vol. 89
MODELING LOVE DYNAMICS

ISBN 978-981-4696-31-9

Printed in Singapore

Preface

This book shows how love stories—an issue of vital concern in our lives—can be tentatively described using mathematical models. As modeling will be a relatively unfamiliar technique to some readers, we first explain in this Preface why have we developed mathematical models and why are they important for scientific research. We then focus on the aim of the book, namely, on obtaining and analyzing some of the most reliable mathematical models available in the literature for describing love stories. These models allow us to formally derive, in agreement with the basic principles of the psychology of love, how love affairs are expected to evolve from the initial state of indifference to the final romantic regime. We end this Preface with a few words about the way the book is organized and with some suggestions for getting the most out of reading it.

A mathematical model is a set of formally stated conjectures regarding the interactions that exist among the variables in a system. A model can be studied, analytically and/or numerically, to predict the evolution of these variables in any hypothetical context. If the predictions are in agreement with the observations obtained in a specific context, then the tendency is to imagine that the conjectures and hence the model itself are valid in all contexts. This is what people mean when they say that a model has been validated. However, it may be that a model gives correct predictions in one context and wrong predictions in others. In other words, validated models can be wrong. Despite having this potentially serious drawback, modeling is pervasive in our life. For example, traffic, weather, energy consumption, and various stock-market indicators are predicted daily using mathematical models, and these forecasts are passed on to the public.

Models can have radically different purposes. They can be developed for making detailed predictions concerning a very particular issue, such as the way a certain suspension bridge oscillates. But they can also be developed for predicting the qualitative behavior of large classes of paradigmatic systems, like thin elastic bars moving in a fluid at constant speed. In the first case the model is different from any other existing model, has many variables (which are needed to describe all the details of interest), and is analytically intractable because of its high complexity. These models are analyzed numerically by computer, which is why they are often called *simulation* models. In the second case, the model describes, in principle, a great variety of similar systems identified by a number of parameters (*e.g.*, the length and the speed of the thin bar), it has only a few variables (because details are few in number), and it is, at least partly, analytically tractable. Models of this kind are called *conceptual* because they explain a given behavioral property of an entire class of systems (for example, the existence, for any length of the bar, of a speed threshold below which the structure does not oscillate). These models are also called *minimal* or *toy* models to show that they involve only a few variables and that they do not claim to quantitatively explain details in too technical a manner.

In any field of science there are plenty of phenomena that are known (from experience, field observations or laboratory experiments) but that have not yet been satisfactorily explained. For example, until the time of Sir Isaac Newton, everybody knew that apples fall from trees, but not why. Newton proposed a satisfactory explanation for the phenomenon, which originated classical mechanics. Thus (in science) knowing why something happens can be more important than knowing that it just happens. We should not, therefore, negatively judge studies (such as those in this book) that do not expand the boundary of known facts, provided that they do expand the boundary of understood facts.

The most frequently used conceptual models are composed of a finite number of Ordinary Differential Equations (ODEs), one for each variable. Interest in these models goes back to the 17th century, when Newton invented differential calculus to present his now-famous law, which allowed him to model the motion of a mass along a line. His model is two-dimensional because there are only two variables: the position and the speed of the mass. Only two centuries later, thanks to the studies of Coulomb and Henry, two ODEs were used in a conceptually similar way to model simple electrical systems involving a voltage and a current. Two-dimensional conceptual models developed to analyze various dynamic phenomena mark the beginning of the modeling era in all fields of science. One century ago, in particular around 1920, there was a spectacular period of scientific innovation with studies on chemical clocks (Bray), predation and competition (Lotka and Volterra), epidemics (Kermack and McKendrick), water pollution (Streeter and Phelps), and electrical oscillators (Van der Pol).

Here we present the conceptual models used in the last decades to describe the development of love affairs. The most simple of them are oversimplified given the complexity of the phenomena characterizing interpersonal relationships. They aim to capture the involvement of an individual by means of a single variable, called *feeling* (or *love*). Thus, the conceptual models of a couple are composed of two ODEs, one for each individual. This extreme simplicity prevents a rich variety of interesting details—discovered on the basis of interviews, self-reports, and psychoanalysis—being included in the model.

The two ODEs are a simple balance between production and consumption flows of love. In accordance with the basic and well established principles of the psychology of love, the production flow is the reaction of an individual to the care expected from the partner. To be more specific, this expectation is split into two components. The first is associated with the involvement of the partner and is responsible for what is called the *reaction to love*, while the second depends only on other traits (beauty, age, social position, wealth, etc.) and gives rise to the *reaction to appeal*. The consumption flow is simply the consequence of the oblivion process— the forgetting of past involvements is necessary to make possible a return to the "gene market" after abandoning or being abandoned by a partner. Oblivion thus has an obvious evolutionary origin. Given that reaction functions depend on love, different classes of individuals with different behaviors can be identified (secure, insecure, unbiased, synergic, platonic). The aim of this book is, in a sense, to show how the evolution of a love story can be virtually predicted from the behavioral characteristics identifying the different categories to which the individuals belong. The identification of micro-macro links of this kind has always been an important issue in the social sciences.

A sensible tradition in scientific books is to specify in the Preface who its interested readers might be. In our case we have three answers. Love being a problem of vital interest for all human beings, the first (and largest) class of potential readers is anyone inclined toward intellectual speculation, as long as he or she has a minimum background in mathematics. To help such people, we present the basic notions involved in mathematical modeling in an Appendix. We also specify at the end of the preface of each chapter if the reader requires any special mathematical skill to understand it.

Some specific case studies are analyzed in detail and described in chapters where almost no mathematics are required for comprehension. To involve the reader, these case studies deal with love stories described in novels and films that are famous worldwide or are found in classical love poetry and prose. Readers without much mathematical knowledge can start with these chapters so that they can appreciate the power of the modeling approach: they can take on the theoretically oriented chapters later if they are interested.

The other two classes of readers comprise scientists working in theoretical psychology (or sociology) and in applied mathematics (or systems analysis). These two types of readers are so different that it is practically impossible to find a style of presentation that is satisfactory for both. We have therefore decided to take an exceptionally straightforward approach in our discussion of the issues concerning romantic relationships, even in our formal mathematical modeling. In other words, we systematically avoid the elegant jargon and the tendency to engage in a deep philosophical discussion of theoretical psychology; at the same time, we avoid the hermetic formalism of mathematics and present results even without reporting their proofs. We hope that in doing this, we will make the book more accessible to everyone, even though some psychologists may disagree with our oversimplified discussions and some mathematicians may find the presentation of some issues somewhat sloppy. We also limit our analysis to the study of standard interpersonal relationships and use only well known mathematical techniques. This will certainly disappoint readers who are more interested in sophisticated issues and models. Hopefully, the satisfaction of such wishes will be possible in the near future.

The book has two parts, with nine chapters in the first part and five chapters in the second. The book begins with a general Introduction and ends with an Appendix on dynamical systems. In the first part stories and models are simple, in the sense that each individual is characterized by a single variable—interest in the partner. This means that simple models cannot be used to mimic love stories between individuals influenced by fluctuating environmental stresses or by a second relevant emotional dimension like artistic inspiration or love for an additional partner. In the second part, the assumptions justifying the use of simple models are relaxed.

In eight of the chapters, we take a theoretical orientation, presenting the main properties of the evolution of love stories between classes of individuals identified by particular psychological traits. In the remaining chapters, we show how a model can be derived from the characters of the individuals involved and how it can be used to predict the evolution of a specific love story. We do this by referencing love stories described in classical poems, novels, plays or famous films, for example, "Gone with the Wind", "Beauty and The Beast", "Pride and Prejudice", "Cyrano de Bergerac", and "Jules et Jim". These case studies reinforce the findings of the theoretical chapters and show the power of the modeling approach. For the reader's enjoyment, short segments of each film are available at home.deib.polimi.it/rinaldi/films.html.

Acknowledgments

Frederic Jones of the University of Wales claimed in the July 1994 issue of The New Scientist that the amorous ups and downs of Francesco Petrarch, the most lovesick poet of all time, can be intuitively understood from the basic notions of nonlinear dynamics. A few days later, during a boat trip in Berlin, S.R. was told of this claim by Gustav Feichtinger of the Technische Universität Wien and became fascinated by the idea of describing love affairs using mathematical models. This book would likely not exist without that pleasant trip.

Back from Berlin, S.R. showed after a few months of work that Petrarch's amorous cycles can be derived from a mathematical model, thus giving strong support to the conjectures of an enthusiastic Frederic Jones. In the same period, A.G. discovered that, in contrast with the predictions of linear models, couples composed of secure individuals can have alternative romantic regimes, a property that is strategically important for understanding complex love stories. Almost at the same time, A.G. and Feichtinger, pointed out that insecurity and synergism are the components of an explosive mix that triggers turbulence in romantic relationships. This discovery motivated F.D. who explored the issue further and in his 1999 master's thesis presented a mathematical interpretation of the romantic ups and downs of Kathe and her two lovers, as described in the novel "Jules et Jim" by Henry-Pierre Roché. In the same study, F.D. discovered the first strange attractor in romantic relationships, thus explaining why triangular love stories are unpredictable even in the extreme case of constant social environments. Numerous and relevant are the constructive criticisms suggested on this issue by students and colleagues, in particular by José-Manuel Rey, Universidad de Madrid. Finally, F.D.R. and P.L., the youngest in the group, focused in the recent years on the consequences of alternative romantic regimes by studying a number of classical love stories, like "Pride and Prejudice", "Beauty and The Beast", "Gone with the Wind", and "Cyrano de Bergerac". Mathematical interpretation of these love stories is certainly not a trivial exercise, and the quality of the results obtained justifies the effort.

Some of the work (including the masterly editing of the book by Kathryn Platzer) was conducted at the International Institute for Applied Systems Analysis (IIASA), an outstanding Research Institute near Vienna located in Schloss Laxenburg—a castle that belonged to the former Austrian imperial family. Cooperation with IIASA was developed within the framework of the Evolution and Ecology Program, led by Ulf Dieckmann. During their visits to Laxenburg the authors had the chance to meet and discuss with several scholars as well as with guests and visitors attracted to Schloss Laxenburg by the scientific reputation of IIASA. A castle balcony overlooking the park where the Empress Elisabeth of Austria (Sisi) reputedly enjoyed the sun was a fitting place for discussing important issues of love dynamics. The painting by Gustav Klimt on the cover of the book is a respectful homage to the "fin de siècle" atmosphere still perceptible in the town of Laxenburg itself.

Contents

Complex models 119

Chapter 1

Can we model love stories?

In this introductory chapter we first recall that the evolution over time of an interpersonal relationship can be described graphically. The idea is a relatively unsophisticated one because it assumes that the interest (feeling) of one person in another can be captured by a single variable. Thus, a love story is represented by two graphs showing the evolution over time of the feelings of each of the lovers. The same love story can be represented more compactly by a single curve showing the simultaneous evolution of the feelings in a two-dimensional space. In the jargon of dynamical systems theory, this curve is called the *trajectory* and starts from the point that represents the initial feelings of the two individuals. A set of trajectories starting from different initial conditions is called the *state portrait* and is a very effective tool for discussing the consequences of various factors influencing the couple (*e.g.*, secret extramarital affairs).

In the study of romantic relationships, state portraits are hard to obtain by direct observation of the individuals; however, they can easily be produced using mathematical models. Thus, we first present a very short survey of the models used in this context in the past, followed by the main general features of the models that we discuss in this book. The latter are conceptually consistent with the general principles developed over the last decades in the study of psychology of love, in particular Bowlby's work which has received by far the most attention in this field. In our models, the time evolution of the involvement of an individual is dictated by the imbalance between consumption, due to oblivion, and regeneration, due to the response of the individual. This brings us naturally to the notions of reaction to appeal and reaction to love, which are the basic components of the models used in this book. These notions allow secure individuals to be formally defined as those with increasing reaction to love and unbiased individuals as those with reaction functions that are uninfluenced by the individual's own involvement.

No particular mathematical knowledge is needed for understanding this chapter.

1.1   Graphical representation of love stories

Attempts to rigorously define love are numerous and can be seen in various fields of science, in particular, psychology, anthropology, sexology, and sociology. In these studies, love is often conceptualized as the integration of a few basic behavioral dimensions, for example, attachment, caregiving, and sex, as discussed by Bowlby (1969) in his account of *attachment theory*, or as intimacy, passion, and commitment, in the theory proposed by Sternberg (1986). Of these basic dimensions, attachment is the one that has received the most attention (Fraley and Shaver, 2000; Mikulincer and Shaver, 2003), because it is also believed to influence the sexual dimension (Schachner and Shaver, 2004). Similarly, a remarkable number of different catalogs of "love styles" can be found; a quite ancient one, "la Carte du Tendre," is described in *Clelia*, a novel written in the 17th century by Madeleine de Scudéry); moreover, a rich series of procedures are suggested for quantitatively evaluating (in suitable scales) the degree of involvement of one person with another, for example, the *Passionate Love Scale* (Hatfield and Sprecher, 1986a) and the *Romantic Belief Scale* (Sprecher and Metts, 1986). Although there is no general agreement across any of these studies, qualitative statements are frequently found, even in everyday language, that reveal the possibility of making comparisons over time (*e.g.*, "I love her more and more") or among individuals (*e.g.*, "He is certainly more involved than me"). The involvement has also a sign (*e.g.*, "I do not love her anymore; actually, I hate her"), zero being the value corresponding to indifference.

If we assume, as proposed by Levinger (1980), that the feeling of one person for another can be measured by a real number, a love story between two individuals (for example, 1=she, 2=he) can be described by two graphs showing the evolution in time (t) of her and his feelings $(x_1(t)$ and $x_2(t))$. If $t = 0$ is the time at which the love story begins and if the two individuals are initially indifferent to one another $(x_1(0) = x_2(0) = 0)$ the two graphs start from the origin of the space (t, x).

In Figure 1.1 three different love stories are represented. The first is

very simple: both individuals are more and more involved as time goes
on and sooner or later—say after a few months— reach a plateau $(\bar{x}_1, \bar{x}_2)$
where they remain for a very long time, ideally forever. The second story
is similar to the first one in the long term, but not at the beginning, where
he is antagonistic $(x_2 < 0)$. Finally, the third pair of graphs represents a
relatively wild narrative in which she has remarkable and recurrent ups and
downs that entrain him in a similar but less pronounced pattern.

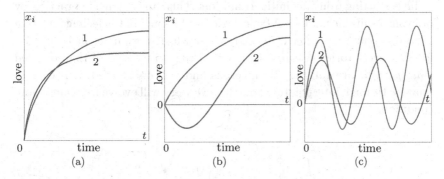

Fig. 1.1 Three love stories starting from the state of indifference $x_1(0) = x_2(0) = 0$.
Her [His] feeling is in red [blue].

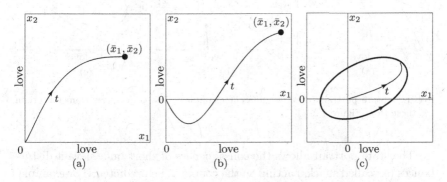

Fig. 1.2 Love stories of Figure 1.1 in the space of the feelings (x_1, x_2).

Each love story can be represented more compactly with a single graph
showing the simultaneous evolution of $x_1(t)$ and $x_2(t)$ in the space of the
feelings. The three panels in Figure 1.2 show the same stories described in
Figure 1.1. The graphs in these figures are called *trajectories* in the jargon
of dynamical systems theory. If 1 and 2 are initially indifferent to each
other, the trajectory starts from the origin at $t = 0$ and develops in time as

indicated by the arrow. The speed at which the point $(x_1(t), x_2(t))$ moves along the trajectory is not constant because the development of love stories is often characterized by accelerations and slowdowns. In particular, when the feelings tend for $t \to \infty$ toward a plateau $(\bar{x}_1, \bar{x}_2)$, as in Figures 1.1a,b the trajectory must slow down when approaching the point $(\bar{x}_1, \bar{x}_2)$ known as the *equilibrium* point.

For various reasons, it may be of interest to describe the evolution of the feelings starting from any initial condition. This can be done, in practice, by drawing a suitable number of trajectories in the space of the feelings starting from different points $(x_1(0), x_2(0))$, as shown, for example, in Figure 1.3 for the three stories already discussed. If the trajectories of this so-called *state portrait* are sufficiently numerous and well distributed, it is easy to imagine through interpolation how the love story will evolve from any initial condition.

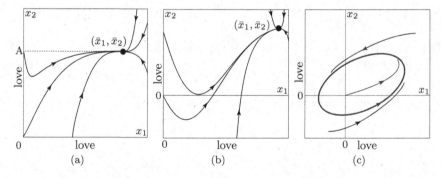

Fig. 1.3   State portraits of three different examples. The trajectories from the origin are those from Figure 1.2.

The state portrait allows the consequences of short but relevant distur- bances (so-called shocks) acting on the couple to be predicted. For example, assume that the couple described in Figure 1.2a has already reached the equilibrium $(\bar{x}_1, \bar{x}_2)$ and that one of the two individuals, say 1, has a tem- porary and secret extramarital affair. As a result, the feeling x_1 drops suddenly, say from $\bar{x}_1$, to 0, while the feeling x_2 remains at $\bar{x}_2$. Thus, once the secret affair is over, the feelings are $(0, \bar{x}_2)$ and not $(\bar{x}_1, \bar{x}_2)$. The state portrait of Figure 1.3 can therefore be used to predict the consequences of the extramarital affair. In the specific case, the new trajectory starting from point A in Figure 1.3a tends again toward point $(\bar{x}_1, \bar{x}_2)$; however, there is a temporary attenuation of the involvement of individual 2. In other words, as far as this couple is concerned, the shock has negative consequences in the short term but is absorbed in the long term. Of course, in other couples, similar shocks can have different consequences, as shown in Figure 1.4 where the shock triggers a transition from a positive romantic regime to a permanent negative (antagonistic) regime.

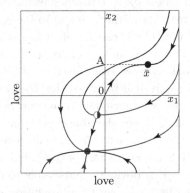

Fig. 1.4 The state portrait of a couple with two alternative stationary regimes. Solid points are stable equilibria, while the half-empty point is a saddle.

State portraits like those shown in Figures 1.3 and 1.4 can easily be obtained in physics and chemistry through continuous and precise measurements (in the laboratory or in the field) of the variables of concern. In other fields, like biology, economics, and sociology (in particular, the study of interpersonal relationships) the observations are quite rare (and often unreliable) samples. In these cases, the trajectories of the state portrait can be obtained only by suitably interpreting the few available samples of information.

1.2 The models

State portraits can also be produced by mathematical models based on the most important characteristics of the couple. At one extreme, these characteristics can be extracted from the analysis of a short discussion between the partners, as done, for example, by Gottman *et al.* (2002a,b) (see also Chapter 5 in (Murray, 2002)). At the other extreme, following the classical approach of psychoanalysis, the characteristics of the couple can be identified from the descriptions of a significant number of events involving the couple, made available by the two partners, by members of their families, or by friends. If we follow these approaches, the model can be developed only for couples that already exist, and this is obviously a serious handicap because the most interesting and challenging problem is to develop models for predicting the potential evolution of love stories of yet unformed couples. Obviously, this ambitious task can be realized only if a model can be built on the basis of psychophysical characteristics (in the following, the way of reacting to the love and to the appeal of the partner) which, being strictly individual, can be identified *a priori*, that is, even in the absence of a partner. If, in accordance with attachment theory, we presume that individual characteristics are mainly formed during the first years of our life, this is equivalent to saying that the evolution of a romantic relationship is, in a sense, predetermined. However, in real life, couples are influenced by numerous and unpredictable events due, in particular, to the turbulence of the social environment. As a result, love stories can deviate, even strongly, from what the characters of the individuals might somehow have predetermined. As it is obviously impossible to model unpredictable events, we assume in this book (with the exception of Chapter 11) that couples live their love story in the absence of relevant exogenous stresses.

Mathematical models capable of producing state portraits like those shown in Figure 1.3 are many in number and have different natures. For example, one could mimic the behavior of the two individuals by splitting their daily life in a series of elementary periods, some of separation, some of family life, and some of intimacy. In these models, sometimes called Individual Based Models (IBMs), the life of the couple can be described virtually in great detail in the hope of obtaining reliable results. However, experiences performed in other fields, like biology and economics, show that the results are quite fragile because they are often sensitive to the particular concatenation of the elementary periods and to their description. It is therefore not surprising that IBMs have never been used to simulate the evolution of love stories.

Different kinds of computer model have been used in a number of studies to predict the evolution of romantic relationships (Baron *et al.*, 1994; Butler, 2011). In the most relevant of these studies (Huesmann and Levinger, 1976) a model, called RELATE, was proposed and extensively used for simulations. Such a model involves recursive manipulations of suitable "dyadic outcome matrices" representing different interpersonal states. It is certainly a dynamic model, but it is much more complex and less elegant than the ODE models described in the next section.

1.3 ODE models

The models used to date to produce state portraits like those shown in Figure 1.3 are based on Ordinary Differential Equations and are therefore known as *ODE models*. These models, introduced by Newton in the 17th century to describe the dynamics of mechanical systems, are composed of a finite number (n) of ordinary differential equations of the first order

$$
\begin{aligned}
dx_1(t)/dt &= f_1(x_1(t), x_2(t), \ldots, x_n(t)) \\
dx_2(t)/dt &= f_2(x_1(t), x_2(t), \ldots, x_n(t)) \\
&\;\vdots \\
dx_n(t)/dt &= f_n(x_1(t), x_2(t), \ldots, x_n(t)),
\end{aligned}
\tag{1.1}
$$

where $x_i(t)$ and $dx_i(t)/dt$, $i = 1, 2, \ldots, n$ are the n relevant variables and their derivatives, and the n functions f_i, $i = 1, 2, \ldots, n$ depend on $x_1, x_2, \ldots, x_n$. In the following, model (1.1) is often written more synthetically as

$$
\dot{x}(t) = f(x(t)),
\tag{1.2}
$$

where $x(t) = (x_1(t), x_2(t), \ldots, x_n(t))$ is known as the *state* of the system and $\dot{x}(t)$ stands for $dx(t)/dt$. Given the initial state $x(0)$, equations (1.1) can be integrated (analytically or, more often, numerically) to produce the evolution of the state, namely $x(t)$ for $t \geq 0$. In reality, equations (1.1) can also be integrated backwards in time, thus producing the past history of the system. A larger class of models is obtained if the first order derivatives in equation (1.1) are formally substituted by fractional order derivatives. Models of this kind are called *fractional order models* and have also been proposed in the context of love dynamics (see, for example, Ahmad and El-Khazali (2007); Koca (2014); Koca and Ozalp (2013); Liu and Chen (2015); Song *et al.* (2010)). However, these models are not considered in this book, as their relevance to the study of love dynamics has not yet been supported by sensible theoretical arguments or sound applications.

In a number of contributions to the literature, it is shown how model (1.1) can be conceptually justified and potentially derived from the general theory of stochastic processes (Bellomo and Carbonaro, 2006, 2008; Carbonaro and Giordano, 2005; Carbonaro and Serra, 2002) or from very general principles, like those of Quantum Mechanics (Bagarello, 2011, 2012; Bagarello and Oliveri, 2010). However, from an operational point of view, these approaches do not seem to be very promising if the aim of the study is to model specific love stories.

Alternatively, model (1.1) can be derived by assuming that the two partners are rational agents maximizing their utility functions. This idea—quite popular in economics and marketing (Fruchter, 2014)— leads naturally to the formulation of a so-called optimal control problem, the solution of which is a set of equations of the kind (1.1). In the study of love affairs, this approach was followed by Hartl and Mehlmann (1984), Jørgensen (1992), and Feichtinger *et al.* (1999), who modeled the famous Italian poet Francesco Petrarch as a rational addict of his desire for Laura. Their analysis gives formal support to the study of Jones (1995), who discovered the turbulence of the romantic relationship between the poet and his mistress (see Chapter 13). More recently, the same approach was successfully followed by Rey (2010, 2013) to find the best trade-off solution between the cost of deterioration of a relationship and the cost needed to keep it alive. The focus is on the phase of marital dissolution and, in fact, Rey's study starts when the feelings of the partners are at their peaks, while nothing is said about the evolution of the love story from the moment of the first encounter. An interesting extension to the case in which marital dissolution may involve a genetic component can be found in Zhou *et al.* (2014), while the influence of society is examined in Bauso *et al.* (2014) using mean-field game models. An algorithm for the numerical solution of Rey's trade-off problem has been developed by Goudon and Lafitte (2015).

The approach we follow throughout this book is definitely less sophisticated. If the variables $x_i(t)$ are thought of as properties characterizing the individuals, equations (1.1) can be interpreted as balance equations in which the functions f_i are the differences between production and consumption flows. The most *simple* models (sometimes called *minimal* or *toy* models) are those with $n = 2$, where $x_1(t)$ and $x_2(t)$ represent the involvement of each individual with the partner. Conversely, models with $n > 2$ are called *complex*. The first part of the book is devoted to simple models, while in the second part we deal with complex love stories involving more than two persons or individuals with multiple emotional dimensions or lovers influenced by the turbulence of their social environment.

It is important to stress that simple models are therefore composed of two ordinary differential equations, one for "she" and one for "he"

$$\dot{x}_1(t) = f_1(x_1(t), x_2(t))$$
$$\dot{x}_2(t) = f_2(x_1(t), x_2(t)),$$

(1.3)

where $x_i(t)$, $i = 1, 2$, is a measure of the love of individual i for the partner j ($j \neq i$). In the following, $x_i(t)$ is also called *involvement* or, more often, *feeling.* Simple models are a crude simplification of reality: first, because love is a complex mix of different feelings (esteem, friendship, sexual satisfaction, etc.) and thus difficult to capture by a single variable; second, because the tensions and emotions involved in the social life of a person cannot be included in a single equation. In other words, in simple models only the interactions between the two individuals are taken into account, with the rest of the world being frozen out and not participating explicitly in the formation of love dynamics.

1.4 Individual characteristics: Oblivion, reaction to appeal, and reaction to love

As already mentioned, the function f_i in model (1.3) is the difference between flows of love that are produced and flows of love that are consumed. In accordance with the principles of attachment theory, the production flows are simply the reactions of individual i to the caring expected from the partner j. Some of these expectations are independent of the feeling x_j of individual j and the reaction to them is here called *reaction to appeal* R_i^A. The remaining expectations are functions of the feeling x_j and the reaction to them is called *reaction to love* R_i^L. The consumption flow is due to the natural process of oblivion and is therefore indicated by O_i. In conclusion, the two right-hand sides of equations (1.3) have the form (Rinaldi and Gragnani, 1998b)

$$f_i = R_i^A + R_i^L - O_i, \qquad (1.4)$$

where the three flows must be further specified.

The oblivion process can easily be studied by looking at the extreme case of an individual who has been abandoned by (or separated from) the partner, which implies $R_i^A = R_i^L = 0$. If we assume, as shown in Figure 1.5, that in such conditions $x_i(t)$ vanishes exponentially at a rate α_i—called the *forgetting* coefficient—we must write

$$\dot{x}_i(t) = -\alpha_i x_i(t), \qquad (1.5)$$

that is to say, the consumption flow is linearly dependent on x_i, that is, $O_i = \alpha_i x_i$. The same linearity assumption is usually made in physics and in chemistry for all consumption flows. But more complex forms for the oblivion flow O_i are also possible, as in the case considered in Chapter 15, where a woman with two lovers is quicker to forget the one she is less involved with. It is worth noting that the forgetting coefficient—undoubtedly

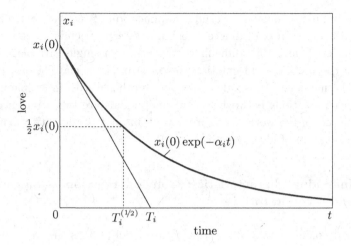

Fig. 1.5   Exponential decay of interest in the partner after separation: $T_i = 1/\alpha_i$ is the time constant and $T_i^{(1/2)}$ is the half-life.

an individual trait—strongly depends upon the culture of the population to which the individual belongs. The solution of equation (1.5),

$$x_i(t) = x_i(0)\exp(-\alpha_i t),$$

is often written in the form

$$x_i(t) = x_i(0)\exp(-t/T_i),$$

where $T_i = 1/\alpha_i$ is called *time constant* and has the simple graphical interpretation shown in Figure 1.5. As $x_i(T_i) = x_i(0)\exp(-1) = 0.37x_i(0)$ we can say that the time constant T_i is a good indicator of the time needed to "practically" forget a former partner. A similar indicator, often used in physics and biology, is the so-called *half-life*, which is the time $t = T_i^{(1/2)}$ at which $x_i(t)$ is one half the initial value $x_i(0)$. Thus,

$$\exp\left(-\frac{T_i^{(1/2)}}{T_i}\right) = \frac{1}{2},$$

from which it follows that

$$T_i^{(1/2)} = T_i \log 2 = 0.69 T_i.$$

As experience shows that the time needed in practice to forget a former partner can range from a few weeks to a few years, we can say that the typical length of love stories in which the oblivion process matters should be in the same range. This means that our simple models cannot claim to describe very short stories involving, for example, a single weekend or very long stories concerning, for example, an entire life.

The appeal $A_{j/i}$ of the individual j perceived by the partner i has various components $A_{j/i}^h$ like physical attractiveness, intelligence, education, age, wealth, and others, that are independent of the feeling x_j. If λ_i^h is the weight that individual i gives to the h-th component of the appeal of the partner, we can define the appeal of j (perceived by i) as

$$A_{j/i} = \sum_h \lambda_i^h A_{j/i}^h.$$

Thus, the appeal of an individual is not an absolute trait of the individual, but rather a value perceived by his/her present or future partner. For brevity, in the following the appeal $A_{j/i}$ will be denoted by A_j.

The flow of interest R_i^A generated in individual i by the appeal of the partner is obtained by multiplying A_j by a factor γ_i identifying the reactiveness of individual i to appeal, that is,

$$R_i^A = \gamma_i A_j = \gamma_i \sum_h \lambda_i^h A_{j/i}^h. \tag{1.6}$$

In many models, all terms in (1.6) are assumed to be time-invariant because in many real love stories the appeals, the weights, and the reactivenesses do not vary significantly in time. In these cases, the reactions to appeal in model (1.4) are simply constant. However, in other cases, the love story must be subdivided into a number of segments, each characterized by distinct values of the reactions to appeal. For example, the famous love story between Scarlett O'Hara and Rhett Butler described in "Gone with the Wind" must be split, as shown in Chapter 7, into two parts, one before and one after the Civil War, because during the war Rhett and Scarlett's situations change: Rhett becomes very rich while Scarlett is forced to face greatly reduced financial circumstances.

The reaction to love R_i^L of individual i, sometimes called *return*, must obviously depend upon x_j. In agreement with the studies on the psychology of love, we subdivide individuals into two classes: *secure* and *insecure* individuals. Secure individuals have positive mental models of themselves and of others (Bartholomew and Horowitz, 1991; Griffin and Bartholomew, 1994) and their romantic relationships are characterized by intimacy, closeness, mutual respect, and involvement. They react positively to their partner's love and are not afraid of someone becoming emotionally close to them. Thus, secure individuals are characterized by increasing reaction

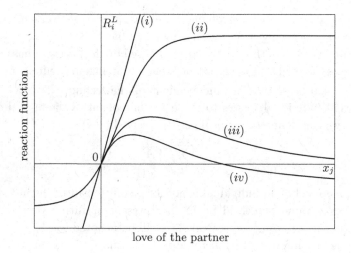

Fig. 1.6 Examples of reaction functions $R_i^L(x_j)$ of secure individuals ((i) and (ii)) and of insecure individuals ((iii) and (iv)).

functions, that is,

$$\frac{dR_i^L(x_j)}{dx_j} > 0.$$

In the extreme case, the reaction R_i^L can be assumed to be linearly increasing with x_j, as shown by curve (i) in Figure 1.6. But, more realistically, $R_i^L(x_j)$ is positive [negative], increasing, concave [convex] and bounded for positive [negative] values of x_j, as shown in Figure 1.6 with curve (ii). The boundedness of the reaction function is a property that interprets the psycho-physical mechanisms that prevent people from reaching dangerously high stress levels. By contrast, insecure individuals, like those indicated with (iii) and (iv) in Figure 1.6, have decreasing reaction functions for values of x_j above a certain threshold. Thus, insecure individuals react less and less strongly at high pressure and involvement because they feel uncomfortable being dependent on and close to others (Schachner and Shaver, 2004).

As is well known, individual reactions can be enhanced by love. For example, mothers often have a biased view of the beauty of their children. This kind of phenomenon, here called *synergism*, was first observed in a study of perception of physical attractiveness (Simpson *et al.*, 1990) which compared individuals involved in romantic relationships with individuals not involved in them. Although we are not aware of any study indicating the

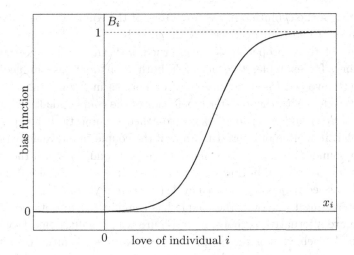

Fig. 1.7   The graph of a typical bias function $B_i(x_i)$.

existence of synergism in the reaction to partner's love, we can reasonably assume that the reaction functions R_i^L can also be enhanced by love. It is also known that individual reactions can be attenuated by love. For example, *platonic* individuals reduce their reaction to the physical appeal of the partner the more they become involved. In order to treat all cases with the same notation and avoid any confusion, we call individuals who are either synergic or platonic *biased*. Thus, the reactions to appeal and to love of biased individuals are written as

$$R_i^A(x_i) = (1 + b_i^A B_i^A(x_i))\gamma_i A_j$$
$$R_i^L(x_i, x_j) = (1 + b_i^L B_i^L(x_i))R_i^L(0, x_j),$$

where $R_i^L(0, x_j)$ is the reaction to love of a completely indifferent individual. The functions $B_i^A(x_i)$ and $B_i^L(x_i)$ are assumed to be zero for $x_i \leq 0$ and increasing from 0 to 1 and first convex and then concave for $x_i > 0$, as shown in Figure 1.7. The constant parameters b_i^A and b_i^L are the bias coefficients and are zero for unbiased individuals and positive for synergic individuals, while b_i^A is negative for platonic individuals.

1.5   From individuals to couples

We have so far discussed the main behavioral characteristics of individuals. If these characteristics are encapsulated in model (1.4), all significant properties of the evolution of the love story of the couple can be extracted from the state portrait produced by the model.

Some of these properties are concerned with the initial phase of the love story, for example, the fact that both individuals are immediately positively involved (as in Figure 1.2a) or not (as in Figure 1.2b). Other properties deal with the long-term behavior of the couple, identified by the limits of $x_1(t)$ and $x_2(t)$ for $t \to \infty$ (so-called asymptotic behaviors). In particular, it is often of interest to know if the couple can have an explosive (*i.e.*, unbounded) behavior for some of the initial conditions or if the couple evolves toward an equilibrium (*stationary regime*) as in Figures 1.2a,b or toward a closed trajectory, called a cycle (*periodic regime*) as in Figure 1.2c. It is interesting to know if the couple has a unique asymptotic regime or if there are alternative regimes, as in Figure 1.4. Finally, some properties refer to the behavior of the couple for a specific set of initial conditions. The most relevant among these properties is *positivity*, namely, the fact that non-negative initial feelings guarantee that future feelings cannot become negative.

The aim of this book is, in a sense, to show how the properties of a couple can be derived from individual properties through the analysis of a mathematical model (this micro-macro link has been discussed by Buder (1991) in the context of dyadic social interactions). We have seen how individuals can be subdivided into secure and insecure and into biased and unbiased, for a total of four classes. If we considered gender differences, we would obtain sixteen types of couples, each with specific properties. If, on the top of this, we wished to distinguish between bounded and unbounded reactions and, among the latter, between linear and nonlinear functions, the final catalog would be so rich as to discourage any attempt at systematic investigation. For this reason, we restrict our analysis mainly to homogamous couples (that is, couples composed of individuals of the same kind), as this is known to be the most frequent case in real life (see Buston and Emlen (2003); Gonzaga *et al.* (2007); Whyte (1990)).

1.6 Summary of results

The book is a mix of theory and applications (case studies). The theoretical chapters require some skill in mathematics, while the applied ones do not. Thus, the reader who is not very familiar with mathematics can start with the case studies and switch to the theory later, possibly after reading the

Appendix on dynamical systems.

The case studies refer to love stories described in well known classical poems or in worldwide famous films. This has two potentially interesting consequences. First, it gives the readers the chance to easily critique our approach and ideas, and, second, it stimulates them to try to suggest more effective alternative models.

In the theoretically oriented chapters the analysis refers to hypothetical couples composed of individuals belonging to specific classes (secure, insecure, unbiased, synergic, platonic). For each class of couples examined, one or more general properties are derived and then illustrated through examples and case studies. Following is a list of the most important properties derived from models, together with an indication of the chapters in which they are discussed. This list can be a useful guide for the reader interested in specific aspects of love dynamics.

- Standard couples (*i.e.*, couples composed of secure and unbiased individuals) have only one stationary romantic regime if the appeal of the individuals is sufficiently high. Moreover, this regime is satisfactory for both individuals (Chapters 2 and 3).
- Standard couples have two alternative romantic regimes if the appeal of the individuals is sufficiently low. The first is favorable because it is positive for both individuals, while the second is unfavorable, because worse than the first and negative for at least one of the individuals (Chapter 3).
- If the individuals of a standard couple with sufficiently high appeals are in love with each other at a given time, they remain so forever (Chapters 2 and 3).
- Two standard individuals with sufficiently high appeal who are completely indifferent to each other when they first meet develop a love story characterized by continuously increasing feelings (Chapters 2 and 3).
- An increase in the appeal of one individual of a standard couple improves the romantic relationship, and the relative improvement is higher for the partner (Chapters 2 and 3).
- Separation and formation of new couples is unlikely in communities of standard individuals if the partner of the n-th most attractive woman is the n-th most attractive man (Chapters 2 and 3).
- Bluffing can be rewarding in love affairs. More precisely, if the individuals of a standard couple are not extremely attractive it makes

sense for them to temporarily give the partner a biased impression of their interest because this helps in reaching the favorable romantic regime. This justifies the behavior certain people exhibit when wooing (Chapter 3).

– Small discoveries can have great consequences in love affairs. More precisely, the discovery of a small hidden component of the appeal can trigger a great transition from an unfavorable to a favorable romantic regime (Chapter 5).

– Couples composed of insecure and unbiased individuals often have two alternative romantic regimes. One of the two regimes is more rewarding for the first individual, while the other is more rewarding for the second (Chapter 6).

– When the regime is reached, the more insecure and/or appealing individual is less satisfied than the partner and is therefore inclined to break off the relationship (Chapter 6).

– Insecurity and bias promote the emergence of recurrent ups and downs in feelings (Chapter 8).

– Exogenous (social and enviromental) stresses can make romantic relationships chaotic, that is, unpredictable (Chapter 11).

– If environmental (exogenous) and romantic (endogenous) clocks tick at comparable frequencies, unpredictability emerges even when stress is light (Chapter 12).

– The existence of an extra emotional dimension, like artistic inspiration, is a key factor that can stabilize or destabilize romantic relationships (Chapter 12).

– Jealousy and conflict make triangular relationships unpredictable (Chapter 14).

– *Ménages à trois* are more unstable than standard triangles composed of one central individual with two separate lovers (Chapter 14).

The above list of properties is by no means exhaustive. The reported properties are only examples that can be theoretically derived through the use of models. In agreement with Vallacher *et al.* (2002), our wish is that the modeling approach described in this book can help researchers discover new aspects of love dynamics or put results that are already known to psychologists into a sound theoretical framework.

PART I
Simple models

In this part, composed of nine chapters, we deal with models in which each individual is characterized by a single variable—the interest of the partner. This means that these models, called *simple*, cannot be used to mimic love stories between individuals who are stressed by their social environments, who have extra emotional dimensions, or who are involved in two or more romantic relationships. All theoretical results are discussed with reference to couples composed of hypothetical individuals, while the examples refer to very specific stories described in well known films—"Cyrano de Bergerac", "Beauty and The Beast", "Pride and Prejudice", "Gone with the Wind", and "Jules et Jim". This gives the reader a chance to better evaluate our approach and propose alternative models for the interpretation of some of the stories.

Chapter 2

Linear models and their properties

In this chapter we analyze the most simple class of couples, namely, that composed of secure and unbiased individuals, sometimes called standard. We here consider the very special case of linear reaction functions, while the general case is studied in the next chapter. The linearity assumption simplifies the study that can be performed analytically using the powerful theory of linear dynamical systems.

After a short historical review of the first contributions in the field (which began forty years ago), five properties are formally derived from the model. These concern the positivity of the couple, the smoothness of the evolution of the love story, its convergence toward a satisfactory stationary regime, and the influence that appeals and reactivenesses have on the quality of the romantic relationship. All these properties are consistent with common wisdom on the problem.

Finally, the above properties are used to discuss the structural stability of a given community, that is, the virtual possibility of non-existence of separation and formation of new couples in the community. The result is that a community is structurally stable if and only if the partner of the n-th most appealing woman is the n-th most appealing man. This result derived from purely theoretical arguments is in agreement with empirical evidence (Hamermesh, 2011; Hatfield and Sprecher, 1986b) and points to the strategic role played by appeal in shaping social communities.

Some knowledge of linear systems theory, in particular the notions of stability, eigenvalues, and eigenvectors (see Appendix), are useful. More details can be found in Rinaldi (1998b).

2.1   Historical premise

The minimal model proposed in the first chapter, namely

$$\dot{x}_i = R_i^A(A_j, x_i) + R_i^L(x_i, x_j) - O_i(x_i), \quad i = 1, 2,$$

is now studied for the simplest class of couples. Indeed, we consider unbiased individuals (*i.e.*, individuals with reaction functions R_i^A and R_i^L depending only on the appeal A_j and on the love x_j of the partner). Moreover, we assume that all reaction functions and oblivions are linear, that is,

$$O_i(x_i) = \alpha_i x_i \qquad R_i^L(x_i, x_j) = \beta_i x_j \qquad R_i^A(A_j, x_i) = \gamma_i A_j,$$

where α_i is the forgetting coefficient and β_i and γ_i are the *reactivenesses* to love and appeal. To avoid cases of limited interest, we also suppose that all parameters α_i, β_i, γ_i and appeals A_i are positive. Note that these assumptions imply that the individuals are secure ($dR_i^L/dx_j > 0$) and have unbounded reaction functions. Thus the model we consider is

$$\begin{aligned} \dot{x}_1 &= -\alpha_1 x_1 + \beta_1 x_2 + \gamma_1 A_2 \\ \dot{x}_2 &= -\alpha_2 x_2 + \beta_2 x_1 + \gamma_2 A_1, \end{aligned} \qquad (2.1)$$

where the four parameters $(\alpha_i, \beta_i, \gamma_i, A_i)$ identify individual i. As far as we know, model (2.1) with $A_i = 0$ is the first model to have been considered in the context of love dynamics. The first attempt to study this model most likely goes back to 1975 when Ilya Prigogine, Nobel Prize winner for Chemistry in 1977, and Etienne Guyon, later Director of Ecole Normale Supérieure, Paris, exchanged some ideas on the subject (a sample of a letter of Guyon to Prigogine is reported in Figure 2.1). Their approach, based on an analogy with thermodynamics, did not produce interesting results and their study remained unpublished. The second study was performed in 1978 at Princeton University by a young student, Steven Strogatz, who wrote an interesting term paper entitled "Differential Equations and the Progress of Love Affairs" for a Sociology exam. The first page of the term paper is reported in Figure 2.2. As explained in a *New York Times* article

http://opinionator.blogs.nytimes.com/2009/05/26/

guest-column-loves-me-loves-me-not-do-the-math,

the term paper was inspired by a romantic relationship in which the student was directly involved. Only ten years later, Strogatz published a one-page paper (Strogatz, 1988) entirely reported in Figure 2.3 and entitled "Love affairs and differential equations" which is now considered the first scientific contribution on the subject. More details, extracted from the term

Phénoménologie du couple. Compléments. Etienne GUYON
Université Paris Sud Orsay

La formulation mathématique ne devrait pas être une cause d'incommu-nicabilité ni la barrière derrière laquelle se protègent les mystificateur ou simplement les faibles — ceux qui ont peur de la rencontre. Dans la clarté et l'universalité de son écriture elle doit, à condition d'être explicitée, avoir un rôle véhiculaire de l'information au même titre, et de façon plus incisive souvent que le mot.

Dans cet esprit je formule le problème énoncé dans les pages précédentes

$$\begin{cases} \dfrac{dv_A}{dt} + \dfrac{v_A}{T_A} + \mathcal{J}t\,v_B = 0 \\[2mm] \dfrac{dv_B}{dt} + \dfrac{v_B}{T_B} + \mathcal{J}t\,v_A = 0. \end{cases}$$

Ceci représente deux équations couplées reliant entre elle la valeur de A, v_A et celle de B, v_B.

Regardons d'abord les 2 premiers termes de la première équation, c'est à dire supposons que la valeur d'échange, d'amour, $\mathcal{J}t$, soit nulle. On a

$$\dfrac{dv_A}{dt} + \dfrac{v_A}{T_A} = 0 \quad \text{ou} \quad \dfrac{dv_A}{dt} = -\dfrac{v_A}{T_A}.$$

$\dfrac{dv_A}{dt}$ représente la vitesse avec laquelle la valeur v_A varie dans le temps. Cette vitesse est d'autant plus grande que T_A, le temps de relaxation de A est court. Cette équation traduit le phénomène d'amortissement dans le temps T_A de A

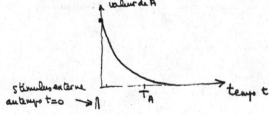

Fig. 2.1 Sample of a letter (in French) of Etienne Guyon to Ilya Prigogine.

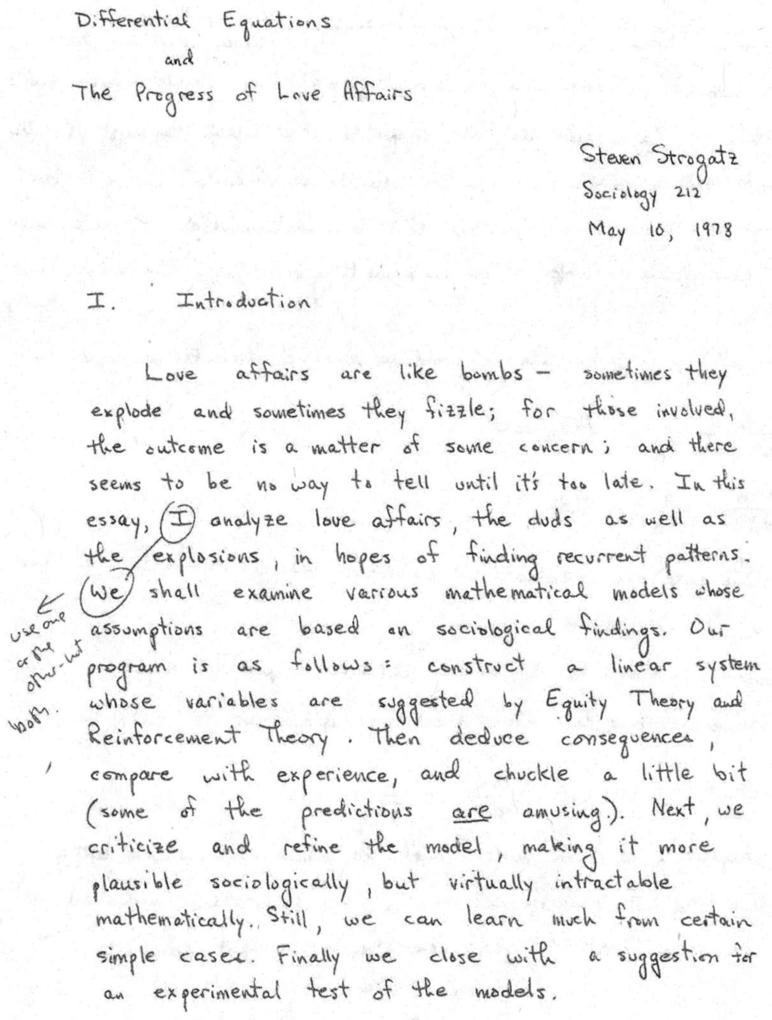

Differential Equations
and
The Progress of Love Affairs

Steven Strogatz
Sociology 212
May 10, 1978

I.   Introduction

Love affairs are like bombs — sometimes they explode and sometimes they fizzle; for those involved, the outcome is a matter of some concern; and there seems to be no way to tell until it's too late. In this essay, (I) analyze love affairs, the duds as well as the explosions, in hopes of finding recurrent patterns. (We) shall examine various mathematical models whose assumptions are based on sociological findings. Our program is as follows: construct a linear system whose variables are suggested by Equity Theory and Reinforcement Theory. Then deduce consequences, compare with experience, and chuckle a little bit (some of the predictions are amusing). Next, we criticize and refine the model, making it more plausible sociologically, but virtually intractable mathematically. Still, we can learn much from certain simple cases. Finally we close with a suggestion for an experimental test of the models.

use one
or the
other—not
both.

Fig. 2.2   The first page of the term paper written by Steven Strogatz for a Sociology exam at Princeton University in 1978.

MATHEMATICS MAGAZINE
VOL. 61, NO. 1, FEBRUARY 1988 35

Love Affairs and Differential Equations

STEVEN H. STROGATZ
Harvard University
Cambridge, MA 02138

The purpose of this note is to suggest an unusual approach to the teaching of some standard material about systems of coupled ordinary differential equations. The approach relates the mathematics to a topic that is already on the minds of many college students: the time-evolution of a love affair between two people. Students seem to enjoy the material, taking an active role in the construction, solution, and interpretation of the equations.

The essence of the idea is contained in the following example.

Juliet is in love with Romeo, but in our version of this story, Romeo is a fickle lover. The more Juliet loves him, the more he begins to dislike her. But when she loses interest, his feelings for her warm up. She, on the other hand, tends to echo him: her love grows when he loves her, and turns to hate when he hates her.

A simple model for their ill-fated romance is

$$dr/dt = -aj, \qquad dj/dt = br,$$

where

$r(t)$ = Romeo's love/hate for Juliet at time t

$j(t)$ = Juliet's love/hate for Romeo at time t.

Positive values of r, j signify love, negative values signify hate. The parameters a, b are positive, to be consistent with the story.

The sad outcome of their affair is, of course, a neverending cycle of love and hate; their governing equations are those of a simple harmonic oscillator. At least they manage to achieve simultaneous love one-quarter of the time.

As one possible variation, the instructor may wish to discuss the more general second-order linear system

$$dr/dt = a_{11}r + a_{12}j$$
$$dj/dt = a_{21}r + a_{22}j,$$

where the parameters a_{ik} ($i, k = 1, 2$) may be either positive or negative. A choice of sign specifies the romantic style. As named by one of my students, the choice a_{11}, $a_{12} > 0$ characterizes an "eager beaver"—someone both excited by his partner's love for him and further spurred on by his own affectionate feelings for her. It is entertaining to name the other three possible styles, and also to contemplate the romantic forecast for the various pairings. For instance, can a cautious lover ($a_{11} < 0$, $a_{12} > 0$) find true love with an eager-beaver?

Additional complications may be introduced in the name of realism or mathematical interest. Nonlinear terms could be included to prevent the possibilities of unbounded passion or disdain. Poets have long suggested that the equations should be nonautonomous ("In the spring, a young man's fancy lightly turns to thoughts of love"—Tennyson). Finally, the term "many-body problem" takes on new meaning in this context.

Fig. 2.3 The one-page paper by Steven Strogatz now considered the first scientific contribution on modeling love dynamics.

paper, were published later by Strogatz in his book (Strogatz, 1994) and also reported (without, however, any relevant additional contribution) in Radzicki (1993) and McDill and Felsager (1994). Strogatz's model has also been used for testing the accuracy of a method of numerical integrations of ODEs (Sunday *et al.*, 2012) and for discussing the dynamic interactions between science and technology (Zhao and Guan, 2013). Extensions of Strogatz's model to the case of interpersonal interactions with time delays have been discussed in Bielczyk *et al.* (2012, 2013). Although these extensions are quite interesting from a mathematical point of view, we do not discuss them here because it is still unclear if they can assist in better interpreting the evolution of love stories. By contrast, more interesting is the extension to the case in which all parameters (including the appeals) are positive (Felmlee and Greenberg, 1999; Rinaldi, 1998b), recently used to extract reliable estimates of physiological parameters from time series of heart and respiration (Ferrer and Helm, 2013). Here we do not report Strogatz's results for two reasons. First, because his model concerns only the case of persons with zero appeal and therefore does not explain how two individuals who are initially indifferent to one another can become involved in a romantic relationship. Second, because one of his individuals reacts negatively to the love of the partner (*i.e.*, $\beta_i < 0$), a rather extreme assumption.

2.2 Five properties of the model

Model (2.1) is a linear system that can be written in the standard form

$$\dot{x} = Ax + b,$$

where the 2x2 matrix A and the vector b are

$$A = \begin{vmatrix} -\alpha_1 & \beta_1 \\ \beta_2 & -\alpha_2 \end{vmatrix} \qquad b = \begin{vmatrix} \gamma_1 A_2 \\ \gamma_2 A_1 \end{vmatrix}$$

The necessary and sufficient condition for asymptotic stability (trace(A) < 0, det(A) > 0) is

$$\beta_1 \beta_2 < \alpha_1 \alpha_2, \tag{2.2}$$

that is, the system is asymptotically stable if and only if the (geometric) mean reactiveness to love ($\sqrt{\beta_1 \beta_2}$) is smaller than the (geometric) mean forgetting coefficient ($\sqrt{\alpha_1 \alpha_2}$). In the following, condition (2.2) is assumed to hold. When this is not the case, namely, when the two individuals are quite reactive to the love of the partner, the instability of the model gives rise to unbounded feelings, a feature which is obviously unrealistic. In that case (*i.e.*, when $\beta_1\beta_2 > \alpha_1\alpha_2$) one must model the couple more carefully by assuming, as is done in the next chapter, that the reaction to love is increasing but bounded with the partner's love.

The matrix A is a so-called Metzler matrix (because it has non-negative off-diagonal elements). This, together with the non-negativity of the components of b, implies the following property.

Property 1. If the individuals are non-antagonistic at a given time, then they remain non-antagonistic forever.

The proof of this property is very simple. In fact, if the property were not true, there would be a trajectory in the space of the feelings (x_1, x_2) leaving the first quadrant by crossing the axis x_i with $\dot{x}_j < 0$. But at that crossing point $x_j = 0$ and $x_i \geq 0$, so that $\dot{x}_j = \beta_j x_i + \gamma_j A_i > 0$, which contradicts the assumption.

As the system is linear and asymptotically stable, $x_i(t)$ tends toward an equilibrium value $\bar{x}_i$, which must be non-negative because the system is positive. But more can be said about the sign of this equilibrium, as specified by the following property.

Property 2. The equilibrium $\bar{x} = (\bar{x}_1, \bar{x}_2)$ of system (2.1) is strictly positive, i.e., $\bar{x}_i > 0$, $i = 1, 2$.

The proof follows immediately (Rinaldi, 1998b) from a general property of positive systems, but can also be obtained by explicitly computing the equilibrium $\bar{x}$, which turns out to be given by

$$\bar{x}_1 = \frac{\alpha_2\gamma_1 A_2 + \beta_1\gamma_2 A_1}{\alpha_1\alpha_2 - \beta_1\beta_2} \qquad \bar{x}_2 = \frac{\alpha_1\gamma_2 A_1 + \beta_2\gamma_1 A_2}{\alpha_1\alpha_2 - \beta_1\beta_2}. \qquad (2.3)$$

Thus, if two individuals meet for the first time at $t = 0$ ($x(0) = 0$), they will develop positive feelings $x_i(t)$ tending toward the positive equilibrium value $\bar{x}_i$. As positive systems have at least one real eigenvalue (the so-called Frobenius eigenvalue, which is the dominant eigenvalue of the system) a second-order system of this kind cannot have complex eigenvalues, that is, the equilibrium of system (2.1) cannot be a focus. In other words, the transients of $x_i(t)$ cannot be damped oscillations characterized by an infinite number of minima and maxima. But even the possibility of a single maximum (minimum) can be excluded, as specified by the following property.

Property 3. Two individuals completely indifferent to each other when they first meet develop a love story characterized by smoothly increasing feelings.

To prove this property consider the null-clines $\dot{x}_i = 0$ (see Appendix), which are straight lines given by

$$x_2 = \frac{\alpha_1}{\beta_1}x_1 - \frac{\gamma_1}{\beta_1}A_2 \qquad (\dot{x}_1 = 0)$$

$$x_2 = \frac{\beta_2}{\alpha_2}x_1 + \frac{\gamma_2}{\alpha_2}A_1 \qquad (\dot{x}_2 = 0).$$

These null-clines (see red and blue lines in Figure 2.4) intersect at point E (representing the strictly positive equilibrium $\bar{x} = (\bar{x}_1, \bar{x}_2)$), and partition the state space into four regions. In the region containing the origin, $\dot{x}_i(t) > 0$, $i = 1, 2$, and this proves the property.

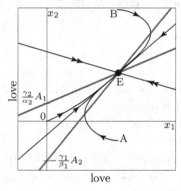

Fig. 2.4 Trajectories and null-clines ($\dot{x}_1 = 0$ in red and $\dot{x}_2 = 0$ in blue) of the system. The straight trajectories are identified by the two eigenvectors. Single and double arrows indicate slow and fast motions.

It should be noted that for non-zero initial conditions, one of the two variables $x_i(t)$ can first decrease and then increase (see trajectory AE in Figure 2.4) or vice versa (see trajectory BE). This can easily be interpreted as follows. Suppose that a couple is at equilibrium and that individual 2 experiences, for some reason, a sudden drop in interest in the partner. The consequence (see trajectory AE in Figure 2.4) is that individual 1 will suffer during the transient that brings the couple back to equilibrium.

Moreover, for particular initial conditions (straight trajectories in Figure 2.4) the two functions $x_i(t) - \bar{x}_i$, $i = 1, 2$, decay exponentially at the same rate (equal to an eigenvalue of the matrix A). The slowest decay occurs along a trajectory that has positive slope and is identified by the dominant eigenvector. On the other hand, the fastest decay occurs along the other straight trajectory which has negative slope. The result is a direct consequence of the well known Frobenius theory (Frobenius, 1912), according to which, in a positive and irreducible system, the dominant eigenvector is strictly positive and there are no other positive eigenvectors (a system is irreducible when it cannot be decomposed into the cascade or parallel connection of two subsystems, a property which is guaranteed in the present case by $\beta_1\beta_2 > 0$). Applied to a second-order system, the Frobenius theory states that the dominant eigenvector has components with the same sign, while the other eigenvector has components of the opposite sign.

We can now focus on the influence of the various parameters on the equilibrium of the couple.

Property 4. An increase in the reactiveness to love [appeal] β_i [γ_i] of individual i gives rise to an increase in the love of both individuals at equilibrium. Moreover, the relative increase $\Delta x/\bar{x}$ is higher for individual i.

The result can be obtained directly from equation (2.3) by deriving $\bar{x}_i$ with respect to β_i [γ_i] and then dividing by $\bar{x}_i$. Condition (2.2), of course, must be taken into account.

It is interesting to note that the first part of the property is a direct consequence of the famous *law of comparative dynamics* (Farina and Rinaldi, 2000). This law states that in a positive system the increase of a positive parameter gives rise to an increase of the components of the state vector at any time, and hence also at equilibrium. The second part of the property is the consequence of a general theorem concerning positive systems, known as the *theorem of maximum relative variation* (Farina and Rinaldi, 2000). This theorem states that if the ith component of the vector b or one element of the i-th row of the matrix A of an asymptotically stable and excitable positive system is slightly increased (in such a way that the system remains asymptotically stable and excitable), the i-th component $\bar{x}_i$ of the state vector at equilibrium is the most sensitive of all in relative terms.

The following, rather intriguing property specifies the influence of appeal on the equilibrium.

Property 5. An increase of the appeal A_i of individual i gives rise to an increase in the love of both individuals at equilibrium. Moreover, the relative increase $\Delta x/\bar{x}$ is higher for the partner of individual i.

The proof comes from Property 4 by noting that in equation (2.3) γ_i is multiplied by A_j, $i \neq j$.

Of course, the reactiveness to love also influences the time needed by the couple to reach the equilibrium. This time is virtually infinite but is practically well identified by the so-called dominant time constant $T_1 = -1/\lambda_1$, where λ_1 is the dominant eigenvalue. But the two eigenvalues λ_1

and λ_2 of the matrix A are the zeros of the characteristic polynomial

$$\det(\lambda I - A) = \det \begin{vmatrix} \lambda + \alpha_1 & -\beta_1 \\ -\beta_2 & \lambda + \alpha_2 \end{vmatrix} = \lambda^2 + (\alpha_1 + \alpha_2)\lambda + \alpha_1\alpha_2 - \beta_1\beta_2,$$

so that the dominant eigenvalue λ_1 is given by

$$\lambda_1 = \frac{1}{2}\left[-(\alpha_1 + \alpha_2) + \sqrt{(\alpha_1 + \alpha_2)^2 - 4(\alpha_1\alpha_2 - \beta_1\beta_2)} \right].$$

Thus, an increase in the reactiveness to love gives rise to an increase in the dominant time constant of the system, which tends to infinity when $\beta_1\beta_2$ approaches $\alpha_1\alpha_2$.

The above properties are easy to interpret. The first states that individuals with positive appeal are capable of establishing a steady romantic relationship. The emotional pattern of two persons falling in love is very regular—beginning with complete indifference, then growing continuously until a plateau is reached. The level of passion characterizing this plateau is higher in couples with higher reactiveness and appeal. Moreover, an increase in the reactiveness of one of the two individuals is more rewarding for the same individual, while an increase in the appeal is more rewarding for the partner (there is a touch of altruism in a woman [man] who tries to improve her [his] appeal). Finally, couples with very high reactiveness are very slow in reaching their plateau. Together with equation (2.3), this means that there is a positive correlation between the time needed to reach the equilibrium and the final quality ($\bar{x}_1$ and $\bar{x}_2$) of the relationship. Thus, passions that develop very quickly (the *coup de foudre*) should be expected to be associated with poor romantic relationships.

Note that the coup the foudre can be particularly violent in the extreme case of unrequited love stories, that is, when one of the two partners is completely insensitive to love (*e.g.*, when $\beta_2 = 0$). In such a case, in fact, the dominant time constant $T_1 = -1/\lambda_1$ is small. This explains, for example, why thousands of teenagers were reported to be in love with Leonardo di Caprio just a few days after the release of the film "Titanic," or why a woman can quickly fall in love with her doctor (at any age).

2.3 Consequences at community level

We can now try to identify the consequences of the above properties at community level. Let us consider a community composed of N women and N men structured in N couples $[A_1^n, \alpha_1^n, \beta_1^n, \gamma_1^n; A_2^n, \alpha_2^n, \beta_2^n, \gamma_2^n]$, $n =$

$1, 2, \ldots, N$, and assume, as usually, that 1 is a woman and 2 is a man. For simplicity, suppose that there are no women (men) with the same appeal, that is, $A_i^h \neq A_i^k$, $\forall(h, k)$ with $h \neq k$. Such a community is considered *unstable* if a woman and a man in two different couples believe it would be to their personal advantage to form a new couple together. In the opposite case the community is *stable*. Thus, practically speaking, unstable communities are those in which the separation and the formation of new couples are quite frequent. Obviously, this definition must be further specified. The most natural way to do this is to assume that individual i would have a real advantage in changing his/her partner if this change is accompanied by an increase in $\bar{x}_i$. However, to forecast the value $\bar{x}_1$ [$\bar{x}_2$] that a woman [man] will reach by forming a couple with a new partner, she [he] should know everything about him [her] (in mathematical terms, she [he] should know his [her] appeal A_2 [A_1] and his [her] behavioral parameters α_2, β_2, and γ_2 [α_1, β_1, and γ_1]). Obviously, this is not the case and the forecast is performed with limited information. Here, we assume that the only available information is the appeal of the potential future partner and that the forecast is performed by imagining that the behavioral parameters of the future partner are the same as those of the present partner. In reality, we should take into account that the appeal has various components. One of them, beauty, is easily recognizable, while some of the others, in particular the moral ones, are hidden and require much more time to be identified. Thus, when we assume that the appeal of the potential future partner is known, we certainly emphasize the role played by beauty in the real process of couple formation.

The above discussion is formally summarized by the following definition.

Definition. A community $[A_1^n, \alpha_1^n, \beta_1^n, \gamma_1^n; A_2^n, \alpha_2^n, \beta_2^n, \gamma_2^n]$, $n = 1, 2, \ldots, N$, is unstable if and only if there exists at least one pair (h, k) of couples such that

$$\bar{x}_1(A_1^h, \alpha_1^h, \beta_1^h, \gamma_1^h; A_2^h, \alpha_2^h, \beta_2^h, \gamma_2^h) > \bar{x}_1(A_1^h, \alpha_1^h, \beta_1^h, \gamma_1^h; A_2^h, \alpha_2^h, \beta_2^h, \gamma_2^h)$$
$$\bar{x}_2(A_1^h, \alpha_1^k, \beta_1^k, \gamma_1^k; A_2^h, \alpha_2^k, \beta_2^k, \gamma_2^k) > \bar{x}_2(A_1^k, \alpha_1^k, \beta_1^k, \gamma_1^k; A_2^k, \alpha_2^k, \beta_2^k, \gamma_2^k),$$

$$(2.4)$$

where the functions $\bar{x}_1(\cdot)$ and $\bar{x}_2(\cdot)$ are given by equation (2.3). *A community which is not unstable is called stable.*

We can now prove that stable communities are characterized by a very simple but peculiar property involving only appeal.

Property 6. A community is stable if and only if the partner of the n-th most attractive woman (n = 1, 2, ..., N) is the n-th most attractive man.

The proof of the property is as follows.

First note that Property 4 implies that condition (2.4) is equivalent to

$$A_2^k > A_2^h, \quad A_1^h > A_1^k, \tag{2.5}$$

that is, a community is unstable if and only if there is at least one pair (h, k) of couples satisfying equation (2.5). Condition (2.5) is illustrated in Figure 2.5a in the space of the appeals where each couple is represented by a point.

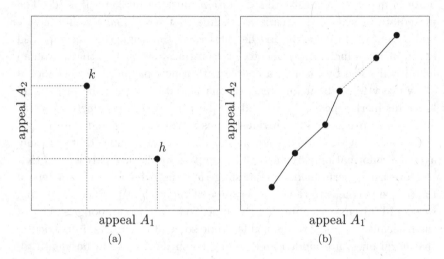

Fig. 2.5 Population structures in the space of the appeals: (a) two points corresponding to two couples (h, k) belonging to an unstable community (see (2.5)); (b) an example of a stable community (each point represents a couple).

Consider a community in which the partner of the n-th most attractive woman is the n-th most attractive man. Such a population is represented in Figure 2.5b, which clearly shows that there is no pair (h, k) of couples satisfying inequalities (2.5). Thus, the community is stable.

On the other hand, consider a stable community and assume that the couples have been numbered in order of the increasing appeal of the women, that is,

$$A_1^1 < A_1^2 < \cdots < A_1^N. \tag{2.6}$$

Then, connect the first point (A_1^1, A_2^1) to the second point (A_1^2, A_2^2) with a segment of a straight line, and the second to the third, and so on until the last point (A_1^N, A_2^N) is reached. Obviously, all connecting segments have a positive slope because, otherwise, there would be a pair of couples satisfying condition (2.5) and the community would be unstable (which would contradict the assumption). Thus, $A_2^1 < A_2^2 < \cdots < A_2^N$. This, together with equation (2.6), states that the partner of the n-th most attractive woman is the n-th most attractive man.

On the basis of Property 6, higher tensions and frictions should be expected in communities where couples are in conflict with the appeal ranking. This result, derived from purely theoretical arguments, is in agreement with empirical evidence as noticed by economists (Hamermesh, 2011) and psychologists (see, for example, Hatfield and Sprecher (1986b), who in the preface of their book explicitly say that "although most people desire attractive partners most often, because of the dynamics of supply and demand, they end up pairing with someone of about their own level of attractiveness"). Indeed, partners with uneven appeal are rarely observed in relatively stable communities. Of course, in making these observations one must keep in mind that appeal is an aggregate measure of many different factors (physical, financial, intellectual, etc.). Thus, for example, the existence of couples composed of a beautiful woman and an unattractive but rich man does not contradict the theory but confirms a classical stereotype.

Chapter 3

Couples composed of secure and unbiased individuals

We continue in this chapter with the analysis of couples composed of secure and unbiased individuals (standard couples) by relaxing, with respect to the previous chapter, the unrealistic assumption of linearity (hence, of unboundedness) of the reaction to love. Moreover, we no longer constrain the appeals to be positive. The main property that makes this model radically different from the linear one is that for suitable values of its parameters it has two alternative stable states. This property can easily be proved by looking at the shape of the two null-clines and by noticing that variations of her [his] appeal shift vertically [horizontally] his [her] null-cline. The result is that in some couples (called robust) the null-clines intersect at a single (globally stable) equilibrium point, while in other couples (called fragile) they intersect at three distinct points, one unstable and two stable. Moreover, one of the two stable states is better than the other, in the sense that its components (the feelings of the two individuals) are greater than the corresponding ones of the other state. This means that one of the alternative stable states is satisfactory while the other is not. A fragile couple can therefore be in a satisfactory or unsatisfactory romantic regime depending upon their past history.

At this point two interesting questions arise. The first question is: which of the fragile couples are converging to the satisfactory romantic regime if the individuals are initially in a state of indifference? The answer can be qualitatively obtained by looking at the null-clines, while a quantitative answer can be derived by performing a relatively simple numerical bifurcation analysis of the model. This and the study of the signs of the feelings at equilibrium allow the complete catalog of behaviors of standard couples to be obtained. The second interesting question is: can the individuals of a fragile couple starting from the state of indifference avoid converging toward the unsatisfactory romantic regime by temporarily applying a suitable behavioral strategy? An answer to this problem can be simply obtained by playing, once more, with the null-clines of the model. The result is that the satisfactory romantic regime is reached if at least one of the two individuals increases through some artifact either the love given to the partner or his/her appeal to the other. Moreover, this bluffing behavior can be interrupted as soon as the state of the couple enters the basin of attraction of the satisfactory romantic regime.

The notions of null-cline, alternative stable state, basin of attraction, and some knowledge of bifurcation analysis (see Appendix) are useful for understanding this chapter. More details can be found in Rinaldi and Gragnani (1998a) and Rinaldi *et al.* (2010, 2015). Theoretical results on the role played by time delays can be found in Liao and Ran (2007).

3.1   The model

We now consider, as in the previous chapter, couples composed of secure and unbiased individuals, but we relax the assumption of linearity (and hence, of unboundedness) of the reaction to love. More precisely, we consider functions $R_i^L(x_j)$ that saturate for large values of the feelings, as shown in Figure 3.1 (see curve (*ii*) in Figure 1.6 of Chapter 1). These functions are positive [negative], increasing [decreasing], and concave [convex] for positive [negative] values of the feeling x_j. A possible example of the analytical expression of these functions is

$$R_i^L(x_j) = \frac{\exp(x_j) - \exp(-x_j)}{\exp(x_j)/R_i^+ - \exp(-x_j)/R_i^-}. \tag{3.1}$$

The two functions represented in Figure 3.1 belong to this class. The *a priori* estimate of the parameters of these functions is undoubtedly a difficult task, although some studies on attachment styles (Carnelly and Janoff-Bulman, 1992; Griffin and Bartholomew, 1994; Levinger, 1980) might suggest ways of identifying categories of individuals with high or low reactions to love. In particular, an accurate estimate of the parameters appears forbiddingly difficult: one would need to sit down with a couple for weeks to obtain data from both partners on a long list of questions (Levinger, 1980). This identification problem is not considered in this chapter (or in the book as a whole), which is centered only on the derivation of the properties of the model.

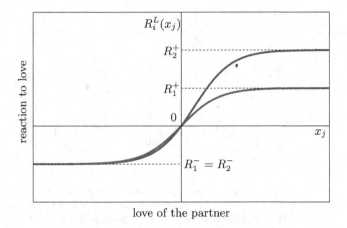

love of the partner

Fig. 3.1 Reactions to love of two secure and unbiased individuals (red for "she "and blue for "he").

In conclusion, the model we study in this chapter is

$$\dot{x}_1 = -\alpha_1 x_1 + R_1^L(x_2) + \gamma_1 A_2$$
$$\dot{x}_2 = -\alpha_2 x_2 + R_2^L(x_1) + \gamma_2 A_1, \qquad (3.2)$$

where the reactions to love R_i^L are of the kind shown in Figure 3.1, the parameters α_i, γ_i are constant and positive, while the appeals A_i are constant but can be negative.

3.2 Alternative stable states

The property of model (3.2) that makes it radically different from the model of linear couples is that for suitable values of its parameters it can have alternative stable states (ASS). The null-cline $\dot{x}_1 = 0$ of model (3.2), that is, the curve in the space (x_1, x_2) where trajectories are vertical, is given by

$$x_1 = \frac{1}{\alpha_1} \Big(R_1^L(x_2) + \gamma_1 A_2 \Big).$$

Similarly, the other null-cline, where trajectories are horizontal ($\dot{x}_2 = 0$), is given by

$$x_2 = \frac{1}{\alpha_2} \Big(R_2^L(x_1) + \gamma_2 A_1 \Big).$$

Generically, the two null-clines intersect at a single equilibrium point or at three equilibrium points, as shown in Figure 3.2. The directions of the

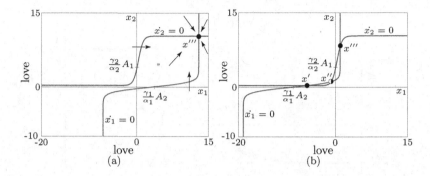

Fig. 3.2   The null-clines $\dot{x}_1 = 0$ (red) and $\dot{x}_2 = 0$ (blue) for R_1^L and R_2^L as in Figure 3.1, $\gamma_1 = 0.5$, $\gamma_2 = 1$, $\alpha_1 = 0.1$, $\alpha_2 = 0.3$, $A_1 = 1.05$, and (a) $A_2 = 0.5$: one equilibrium x''', and (b) $A_2 = -1.9$: three equilibria x', x'', and x'''. Solid points are stable equilibria, while the half-empty point is a saddle.

trajectories in Figure 3.2a indicate that the unique equilibrium point x''' is a stable node, actually a globally stable node. This means that in couples of this kind, feelings tend toward a positive plateau, no matter what the initial conditions are. This is shown in the state portrait in Figure 3.3a. In

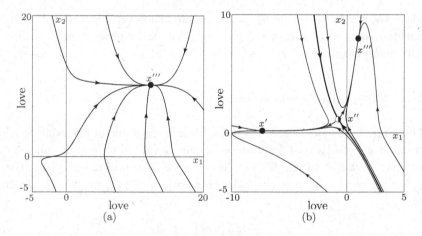

Fig. 3.3   Trajectories in the space of the feelings: (a) parameter values as in Figure 3.2a; (b) parameter values as in Figure 3.2b.

contrast, in the case of Figure 3.2b the directions of the trajectories (not shown) indicate that two equilibria (x' and x''' in the figure) are stable nodes while the other (x'') is a saddle. The geometry of the null-clines also implies that the three equilibria are in the order of satisfaction as

$x' < x'' < x'''$. Obviously, the basins of attraction of the two stable nodes are separated by the stable manifold of the saddle x'' as shown in the state portrait of Figure 3.3b.

Properties 1–6, proved in Chapter 2 for linear couples hold true also for model (3.2) provided the appeals are positive and sufficiently high (in which case the equilibrium is unique, as in Figure 3.2a). The six properties are simple consequences of the geometry of the null-clines. For example, Property 1—the strict positivity of the unique equilibrium x''' in Figure 3.2a— obviously follows from the positivity of the appeals. Thus, the proofs of these properties are not given here. The interested reader may refer to Rinaldi and Gragnani (1998a).

3.3 Bifurcation analysis

Varying the appeal A_1 [A_2], that is, shifting the null-cline $\dot{x}_2 = 0$ [$\dot{x}_1 = 0$] vertically [horizontally], the equilibria x' and x'' in Figure 3.2b can collide and disappear and the same can occur to the equilibria x'' and x'''. These collisions are known as saddle-node bifurcations (see Appendix). Thus, for intermediate values of the appeals there are two alternative stable states (see Figure 3.3b) while for large values of the appeals the equilibrium is unique and globally stable (see Figure 3.3a). The region in the space of the appeals giving rise to alternative stable states can be produced through numerical bifurcation analysis and continuation (Rinaldi *et al.*, 2010). Such a region, indicated by ASS, is shown in Figure 3.4 and has $x'' = x'''$ on its lower boundary and $x' = x''$ on its upper boundary. The signs of the feelings at equilibrium are therefore uniquely determined outside region ASS as indicated in Figure 3.4.

In order to determine the full catalog of behaviors of standard couples, we partition the set ASS in subsets (i) characterized by the signs of the feelings of the two alternative stable states x' and x'''. Thus, any subset (i) is identified by a pair of serial numbers, ranging from I to IV, indicating in which quadrant x' and x''' fall. For example, the case of Figure 3.3b corresponds to the subset (II,I) because x' is in the second quadrant while x''' is in the first quadrant. We must also indicate for each subset (i) which of the two equilibria has the basin of attraction containing the origin. This is simply done by writing the corresponding serial number in bold. For example, in the case of Figure 3.3b the subset is (**II**,I) because starting from the origin, that is, from the state of indifference, the couple evolves toward

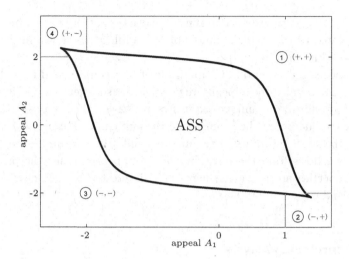

Fig. 3.4   Region of Alternative Stable States (ASS) in the space of the appeals and signs of the feelings in regions ①, ②, ③, and ④ outside ASS. Parameters and functions as in Figures 3.1 and 3.2.

the unsatisfactory state x'. Thus, (II,**I**) and (**II**,I) are different subsets of the partition of ASS. The partition can easily be determined through continuation algorithms (Rinaldi *et al.*, 2010) and the result is shown in Figure 3.5. Couples with (A_1, A_2) in the subsets ⑫, ⑬, ⑭ evolve from the state of indifference toward a positive state x''' because their second cardinal number is **I**. Conversely, couples with (A_1, A_2) in the subsets ⑥, ⑧, ⑩ in which the second cardinal number is I (gray region in Figure 3.5) converge to x' from the state of indifference and to x''' only if the initial feelings are sufficiently high, as explicitly shown in Figure 3.3b.

Now that we have the complete catalog of the possible behaviors of standard couples (Figure 3.5) it would be nice to have one convincing and well documented example for each possible behavior in our theoretically derived catalog. Examples could be taken from the technical literature, but in a sense they would be more appealing if they could be taken from the overwhelming number of popular novels, poems, films, and songs dedicated to love stories. Examples for the largest classes of couples, like those of robust couples, immediately come to mind. "Titanic," a successful 1997 film starring Leonardo Di Caprio (Jack) and Kate Winslet (Rose), is a good example of a robust and high quality couple (region ① in Figure 3.5). It describes the love story that develops on the ship, the Titanic, between Jack and Rose while crossing the Atlantic Ocean. The story starts at the

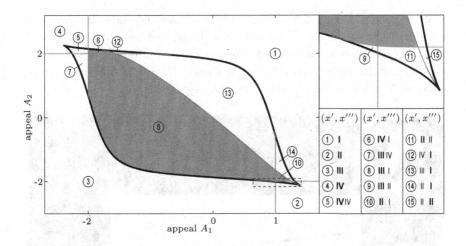

Fig. 3.5 Partition of ASS into subregions ⓘ, $i = 5, \ldots, 15$. In each subregion the two alternative stable states x' and x''' lie in the two quadrants reported in the legend, where the state reached from the state of indifference is in bold. The gray region ⑥∪⑧∪⑩ is the bluffing region (see next section).

port of departure and ends only a few days later because Jack dies when the Titanic sinks after colliding with an iceberg. The two young lovers are so appealing and the level of intimacy they reach in a few days is so high that one can only imagine that this love story would end (if not dramatically interrupted) in a positive equilibrium. Actually, the stability of this equilibrium is also shown in the film, because Rose quickly returns to Jack after a short hesitation caused by the actions of a jealous fiancé.

"Fatal Attraction", a 1987 film starring Michael Douglas (Dan) and Glenn Close (Alex) is a good example of a robust couple with unbalanced feelings (region ④ in Figure 3.5). Indeed, while Dan and Alex are both physically very attractive, Alex is severely mentally ill. The result is that Dan has a positive appeal for her, while Alex has a negative appeal for him. In agreement with our theory, the relationship evolves toward an unbalanced state that ends in tragedy.

Thus, in conclusion, "Titanic" and "Fatal Attraction" are films that could be associated with regions ① and ④ of our bifurcation diagram. Figure 3.6 is far from being complete and the reader is invited to propose films that can be associated with the still empty regions. In doing this, the reader should carefully exclude love stories of individuals with appeals that vary so much during the story that they sequentially involve two or more regions of Figure 3.5. Indeed, in such cases the entire love story is more properly viewed as the concatenation of different love stories separated by the events (a war, an accident, the discovery of a hidden aspect of the character of the partner, etc.) that have caused the variation of the appeals. Chapter 5 is devoted to the study of such cases.

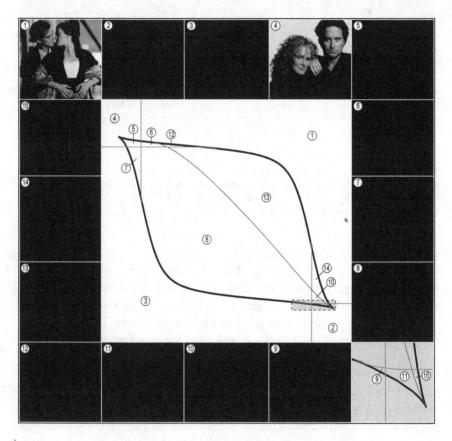

Fig. 3.6   Same as Figure 3.5 with two examples.

3.4   Temporary bluffing in love affairs

Individuals involved in romantic relationships often modify their behaviors intentionally to give the partner a biased impression of their interest. This bluffing behavior can be temporary, as it often is when wooing, or permanent, as in many computer-mediated ("chatting") relationships. Individuals can also bluff by increasing their appeal through some suitable hidden artifact. This form of bluffing can also be temporary, as in the story described in the next chapter, or permanent, as in the case of cosmetic surgery.

The extent of bluffing in our societies is so relevant that one might be led to conjecture that for one reason or another bluffing is often rewarding in love affairs. Here formal support is given to this conjecture through a study of the dynamic behavior of standard couples. Our bifurcation analysis has shown that couples with intermediate appeals (more precisely, couples in the gray region of Figure 3.5), evolve toward an unsatisfactory romantic regime if they start from the state of indifference. However, those couples would converge toward a satisfactory regime if their initial conditions could be suitably modified.

The idea of modifying the initial values of the feelings with a suitable hidden artifact then immediately comes to mind. The idea can be realized through *temporary bluffing*, that is, by increasing for a certain time the appeals from A_i to

$$A_i^* = A_i + B_i,$$

where B_i is a measure of the bluffing intensity of individual i. This form of bluffing has the effect of shifting the two null-clines. More precisely, if individual 1 [2] is bluffing (*i.e.*, if A_1 [A_2] is increased) the null-cline $\dot{x}_2 = 0$ [$\dot{x}_1 = 0$] in Figure 3.2b moves upward [rightward]. Thus, if bluffing is sufficiently strong, the state portrait corresponding to the new temporary appeals (A_1^*, A_2^*) is as in Figure 3.3a, where the feelings can only evolve toward a positive state x'''. Once state x''' has been approached, that is, once a state $\tilde{x}$ close to x''' has been reached, bluffing can be interrupted. Indeed, starting from state $\tilde{x}$ with the real appeals (A_1, A_2) the feelings will evolve as in Figure 3.3b along a trajectory tending toward the positive state x'''.

A second form of bluffing, actually more common than the first, takes place when individual i systematically modifies (intentionally or not) his/her behavior and gives the partner j a biased impression of his/her involvement. In this way, the reaction of j to the love of i in model (3.2) is not $R_j^L(x_i)$, as it should be, but $R_j^L(x_i^*)$, where

$$x_i^* = x_i + B_i.$$

Thus, the two null-clines become

$$x_1 = \frac{\gamma_1 A_2}{\alpha_1} + \frac{R_1^L(x_2 + B_2)}{\alpha_1} \qquad x_2 = \frac{\gamma_2 A_1}{\alpha_2} + \frac{R_2^L(x_1 + B_1)}{\alpha_2}.$$

If individual 1 [2] is bluffing (*i.e.*, if x_1 [x_2] is increased by an amount B_1 [B_2]) the null-cline $\dot{x}_2 = 0$ [$\dot{x}_1 = 0$] shifts leftward [downward] of an amount B_1 [B_2]. Thus, the two null-clines do not move in the same way as in the previous case, but the final result is unchanged. Indeed, if bluffing is sufficiently strong, the two null-clines intersect only at a single point, as in Figure 3.3a, and the feelings evolve toward a positive regime. Interrupting the bluffing when that positive regime has been sufficiently approached, allows the couple to evolve naturally, as in Figure 3.3b, toward the desired satisfactory regime. Of course, the same result is obtained if the two forms of bluffing take place at the same time.

We can complement what we have found on temporary bluffing with the following remarks.

- Temporary bluffing does not need to start at the beginning of the relationship, as it does in computer-mediated romantic relationships (chatting). However, it is difficult to imagine how bluffing could be activated in an already established romantic relationship.
- Temporary bluffing can be unilateral without the chances of success being lowered, because the geometry of the gray region in Figure 3.5, called the bluffing region, guarantees that it is always possible to reach points in region ① moving horizontally or vertically from any point of the bluffing region.
- Temporary bluffing is not needed in couples composed of appealing individuals because such couples always converge to the satisfactory regime (at least one of the two individuals has negative appeal in the bluffing region).
- Temporary bluffing is of no interest for couples with very low appeals, *i.e.*, for couples in regions ②, ③, ④ of Figure 3.5, because when bluffing is interrupted these couples tend inexorably toward a non-satisfactory regime. This means that bluffing makes sense for these couples only if it is adopted permanently (Ben-Ze'ev, 2004).
- In real life the geometry of the bluffing region and the appeals of the individuals are perceived with great uncertainty. It is therefore not surprising (because it is conceptually consistent) that people with low self-esteem (who underestimate their appeal) have a higher tendency to bluff when wooing.

Chapter 4

Roxane and Cyrano

In this chapter a famous love story is analyzed using a mathematical model, with the aim of showing the power of temporary bluffing pointed out in the previous chapter.

The story is described in "Cyrano de Bergerac," a heroic comedy written in verse by Edmond Rostand (1897), a French neo-romantic poet and dramatist. The first performance of the play (Theatre de la Porte Saint-Martin, Paris, 28 December 1897) was a triumph and soon afterwards the play was translated in many languages. Nowadays, "Cyrano de Bergerac" is considered a masterpiece of the French literature on love. "Cyrano de

Gérard Depardieu   Anne Brochet   Vincent Perez

Cyrano de Bergerac   Roxane   Christian de Neuvillette

Fig. 4.1   Shots of the film "Cyrano de Bergerac".

Bergerac" has inspired a number of films, the most successful being directed in 1990 by Jean-Paul Rappeneau, and starring Anne Brochet as Roxane, Gérard Depardieu as Cyrano, and Vincent Perez as Christian (see Figure 4.1). The film is in verse and follows the original play closely.

No new mathematical prerequisites are needed for reading this chapter. Some extra detail can be found in Rinaldi *et al.* (2015).

4.1 The story of Roxane and Cyrano

To support the mathematical interpretation of the love story between Roxane and Cyrano we have isolated eight segments of the film (CB1,...,CB8). These segments and the text of the play (in French) and its translation in English are available at

`home.deib.polimi.it/rinaldi/CyranoDeBergerac`

Still frames of the segments are shown in the panels of Figure 4.2, with the initial and final times of each segment being reported in the corresponding panel.

The story takes place in Paris, where Cyrano de Bergerac—a brilliant poet and swordsman—finds himself strongly attracted to his beautiful cousin Roxane. Although both clever and charismatic, Cyrano, who has a gigantic nose, considers himself too ugly to risk confessing his feelings to Roxane. Roxane one day confides to Cyrano that she is attracted to Christian, a young good-looking Cadet of Gascoyne (CB1). When Cyrano meets Christian (CB2) he embraces him and tells him about Roxane's feelings. Delighted at first, Christian then becomes distraught because he considers Roxane to be an intellectual while he is a simple, unpoetic man. Then, Cyrano has a brilliant idea: he can write Roxane love letters signed by Christian. The latter is happy to agree, welcoming the opportunity to reach Roxane's heart through this bluffing strategy. One day, Christian decides he no longer wants Cyrano's help, and then makes a fool of himself trying to speak seductively to Roxane (CB3). Unfortunately, all he can come up with is "I love you". Roxane is disappointed and sends Christian away. Promptly, under cover of darkness, Cyrano makes Christian stand in front of Roxane's balcony and speak to her, while he stands hidden under the balcony, whispering to him what to say (CB4). At the end of this bluffing ballet, Cyrano forces Christian to hop on to the balcony where Roxane kisses him for the first time (CB5). Soon after that, Roxane and Christian get married but are immediately separated because the Cadets of Gascoyne are sent to the front lines. One day, at the front, Christian guesses Cyrano's secret feelings for Roxane and asks him to find out which

Fig. 4.2   Still frames of the eight film segments (CB1,...,CB8) described in the text.

of the two she would choose. When Cyrano meets Roxane she tells him that she is deeply involved with Christian because of his letters and would love him even if he were ugly (CB6). This is a clear indication of the success of Cyrano's bluffing. About to reveal his secret, Cyrano is interrupted by a gunshot that mortally wounds Christian. Roxane, still ignorant of Cyrano's feelings (CB7), immediately disappears. Thus, Cyrano misses the chance of transforming his bluffing into a successful love story.

Fifteen years later, Cyrano discovers that Roxane lives in a convent and visits her every week. One day he appears at the convent limping and distressed having been wounded in an ambush. As night falls, Cyrano asks to read Christian's last letter to her. He reads it, and when it is completely dark he continues to read, thus revealing to her that he knows the letter by heart. This is how she discovers that the man she has loved during her entire life is right in front of her (CB8) and the bluffing finally stops. Roxane exclaims that she loves him, but Cyrano collapses and dies in her arms. In other words, for the second time Cyrano misses the chance of transforming his bluffing into a successful love story.

4.2   The model of Roxane and Cyrano

In this section we show that the love story between Roxane and Cyrano can satisfactorily be interpreted using the model discussed in the previous chapter, provided the parameters are fixed at suitable values. As quantitative data on the love story do not exist, we cannot use standard identification procedures for tuning the parameters. Instead, we are forced to fix them on the basis of purely subjective impressions. Assuming that 1 is Roxane and 2 is Cyrano, our choice is

$$
\begin{array}{llllll}
\text{Roxane} & \alpha_1 = 0.1 & \gamma_1 = 0.5 & A_1 = 1.05 & R_1^+ = 1 & R_1^- = -1 \\
\text{Cyrano} & \alpha_2 = 0.3 & \gamma_2 = 1 & A_2 = -1.9 & R_2^+ = 2 & R_2^- = -1.
\end{array}
$$

The parameter R_i^- has been given the same values for the two individuals because even subjective impressions are missing for that parameter. Quite low values have been assigned to the forgetting coefficients α_i because both Roxane and Cyrano keep the memory of their love for a very long period (fifteen years) of separation (CB8). Roxane's forgetting coefficient is lower than that of Cyrano because, as may be imagined, being confined in a convent she has not had much of an opportunity to forget her lover, while Cyrano has had a better chance of doing so by remaining actively involved in the social life of the Cadets of Gascoyne. The impression gained when reading Cyrano's wildly enthusiastic proposals is that the maximum reaction R_2^+ and the sensitivity γ_2 of Cyrano are higher than those of Roxane. Finally, the two appeals are of opposite sign in view of the contrast between the beauty of Roxane and the deformity of Cyrano (see Figure 4.1). The appeals A_1 and A_2 can be conveniently compared with Christian's appeal A_2^* which is assumed to be equal to 0.5.

The story can be subdivided into two real phases, indicated by (b) and (d). Phase (b) starts with the agreement between Cyrano and Christian (CB2) and ends with Christian's death (CB7). It is a bluffing phase because Cyrano's soul is hidden in the body of Christian. In other words, Cyrano by some artifact succeeds in increasing his appeal from A_2 to A_2^*. Phase (d) goes from the death of Christian to Roxane's discovery of the truth (CB8) and is simply a phase of separation where oblivion is the only active process.

For the sake of clarity, we also consider three virtual phases indicated by (a), (c), and (e), respectively. In phase (a) Christian does not exist and Cyrano reveals his interest to Roxane. At the beginning of phase (c), just after the death of Christian, Cyrano reveals the truth to Roxane and the feelings of the two lovers evolve toward an equilibrium. This is a virtual phase of no bluffing that should end in a positive equilibrium if the previous bluffing phase (b) was successful. In phase (e) Cyrano remains alive after Roxane has discovered the truth in the convent.

The results obtained by applying the model in each one of the five phases are discussed below.

- *Phase* (a). For the parameter values we have proposed, the two null-clines $\dot{x}_1 = 0$ and $\dot{x}_2 = 0$ are as in Figures 4.3 and correspond to those in Figure 3.2b in Chapter 3. The appeals (A_1, A_2) of the couple fall into subregion ⑩ of the bluffing region presented in the previous chapter. If Cyrano revealed his feelings to Roxane, their story would evolve along trajectory a in Figure 4.3, ending at point A, which is the unsatisfactory state x' of Figure 3.2b. Thus, Cyrano is right in hiding his feelings and in uniting with Christian to initiate a more promising phase of temporary bluffing.
- *Phase* (b). The null-clines are obtained from those in Figure 4.3 by simply shifting the first one to the right by an amount that is proportional to the bluffing $B_2 = A_2^* - A_2 = 2.4$ used by Cyrano. The

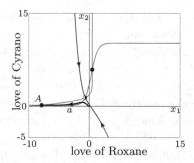

Fig. 4.3   Phase (*a*): Roxane and Cyrano (without Christian).

resulting null-clines corresponds to those in Figure 3.2a in Chapter 3 and intersect only at a single point x''' in the first quadrant. This point is indicated by B in the following. Thus, the feelings evolve along the trajectory b in Figure 4.4 starting from the origin and ending close to point B when Christian dies (CB7). The positivity of the equilibrium B proves that this bluffing phase was successful.

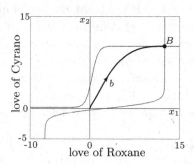

Fig. 4.4   Phase (*b*): Roxane and Cyrano (with Christian).

- *Phase* (*c*). This no-bluffing phase, corresponding to trajectory c in Figure 4.5, is virtual because it would be realized only if Roxane had discovered the truth at Christian's death. Thus, the trajectory c starts from B, develops in agreement with the model equations where $A_2 = -1.9$ is the appeal of Cyrano (therefore the null-clines are those in Figure 4.3), and ends at the equilibrium point C that coincides with x''' in Figure 3.2b of Chapter 3.
- *Phase* (*d*). During this phase—starting at Christian's death—there are no reactions to love and appeal as the two lovers are separated.

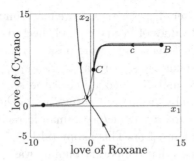

Fig. 4.5   Phase (c): Roxane discovers the truth at Christian's death.

Thus, the couple is described by the reduced model

$$\dot{x}_1(t) = -\alpha_1 x_1(t)$$
$$\dot{x}_2(t) = -\alpha_2 x_2(t),$$

which implies that both feelings systematically decrease (see trajectory d in Figure 4.6 starting from B and tending toward the origin). This evolution is very slow because the forgetting coefficients are low. Thus, when the truth is accidentally discovered by Roxane (CB8) the feelings of the two lovers are still relatively high, as shown by the terminal point D in Figure 4.6.

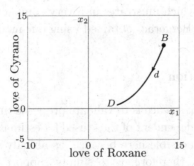

Fig. 4.6   Phase (d): Roxane in convent.

– *Phase (e)*. As in this virtual phase Cyrano remains alive and responds to Roxane's love and appeal, the model is again that used in phase (a), but with initial conditions represented by point D. However, one can reasonably suspect that the sensitivities γ_1 and γ_2 of the two lovers to the appeal of the partner are now lower than those they had when they were younger. But no suggestions emerge from Rostand's

play on this matter. For this reason, we have simulated the model for
various reduced values of γ_1 and γ_2 and checked that in all cases the
feelings of the two lovers tend toward positive values. Figure 4.7 shows
three of these simulations (see trajectories e_1, e_2, and e_3) obtained
for a 25%, 50%, and 75% reduction of the sensitivities γ_1 and γ_2.
Thus, Roxane and Cyrano tend naturally, that is, without the need
for bluffing, toward a satisfactory romantic regime (see points E_1,
E_2, and E_3 in Figure 4.7). This happy (although virtual) end of the
overall story shows that temporary bluffing was successful in the love
story between Roxane and Cyrano.

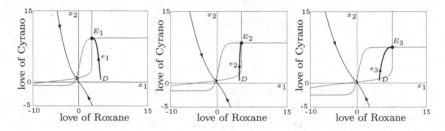

Fig. 4.7   Phase (e): Roxane and Cyrano, if Cyrano had not died.

The results obtained from the analysis of the five phases (a)–(e) are collected in Figure 4.8 which illustrates in a very compact form the interpretation of "Cyrano de Bergerac" obtained using our model.

4.3   Model validation

A sensible tradition in science is to validate proposed mathematical models
using data that are independent of those used for estimating the parameters.
Of course, this is impossible here because, as already said, no data exist.
The best we can do, therefore, is to simply support the model through
highly qualitative arguments, which are, however, independent from those
used to suggest the parameter values. Although we have only two such
independent arguments, both actually weak, we believe that this can be
considered a remarkably lucky case in a field like love dynamics.

The first argument is concerned with the fact that our model supports
the idea that temporary bluffing was potentially successful. But Rostand's
play, ending with Cyrano's death, obviously cannot say anything on this
issue, even if Roxane reveals a definite interest in the future when she

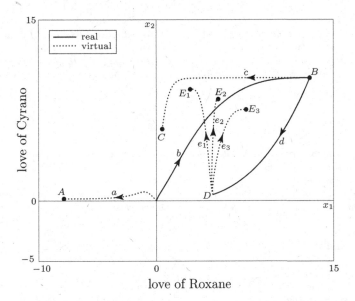

Fig. 4.8 Trajectories of the feelings in each phase (a)–(e) described in the text: (a) virtual no bluffing phase; (b) bluffing phase; (c) virtual phase of no bluffing after Christian's death; (d) separation phase; (e) virtual phase of no bluffing after Roxane's discovery of the truth if Cyrano did not die. The three trajectories e_1, e_2, and e_3 are obtained with sensitivities γ_1 and γ_2 reduced by 25%, 50%, and 75%.

exclaims in the last scene that Cyrano cannot die. We must therefore rely on the impression that the majority of those who have read the play or have seen the film agree with our model's conclusion. Actually, a sort of confirmation of this impression can be found in another film inspired by Rostand's play, entitled "Roxanne" and directed in 1987 by Fred Schepisi, starring Daryl Hannah as Roxane and Steve Martin as Cyrano. The film is a free transposition of "Cyrano de Bergerac" into modern times (Cyrano is the chief of the fire unit in a small American town while Christian is a simple fireman in the same unit), and is aimed at a large and not particularly sophisticated audience. In that film, in order to avoid missing the message on the effectiveness of temporary bluffing, Cyrano does not die and the story ends with Roxane kissing him despite his nose.

The second supporting argument is the agreement between the predictions of the model and the following poignant statement made by Roxane (Scene 4.VIII).

Roxane:

Je viens, ô mon Christian, mon maître!	*O, Christian, my true lord, I come—*
Je viens te demander pardon.	*—I come to crave your pardon.*
De t'avoir fait d'abord, dans ma frivolité,	*For the insult done to you when, frivolous,*
L'insulte de t'aimer pour ta seule beauté!	*At first I loved you only for your face!*

Christian:

Ah! Roxane!	*Roxane!*

Roxane:

Et plus tard, mon ami, moins frivole—	*And later, love—less frivolous—*
Oiseau qui saute avant tout	*Like a bird that spreads its wings,*
á fait qu'il s'envole,	*but cannot fly*
Ta beauté m'arrêtant, ton âme m'entraînant,	*Arrested by your beauty, by your soul*
Je t'aimais pour les deux ensemble!	*Drawn close—I loved for both at once!*

Christian:

Et maintenant?	*And now?*

Roxane:

Eh bien! toi-même enfin	*Ah! you yourself*
l'emporte sur toi-même,	*have triumphed o'er yourself,*
Et ce n'est plus que pour ton âme que je t'aime!	*And now, I love you only for your soul!*

Roxane's statement contains two messages: the first is that her initial involvement was solely a reaction to Christian's good looks. This is in agreement with the model because at the beginning of the story, when $t = 0$, we have $x_1(0) = x_2(0) = 0$ and $R_1^L(0) = R_2^L(0) = 0$, so that

$$\dot{x}_1(0) = \gamma_1 A_2^*.$$

That is to say, Roxane starts to be involved, because A_2^*—the appeal of Christian—is positive. However, it is fair to say that this agreement is not very significant, because condition $R_1^L(0) = R_2^L(0) = 0$ guarantees the same conclusion for any choice of the reaction functions. More interesting is the second message contained in Roxane's statement, namely, that at the end of phase (b) the soul of Christian is for her the dominant source of love. This can be compared with the prediction of the model. In fact, at the end of phase (b), just before Christian's death, we are close to the equilibrium point

$$B = (x_1''', x_2''') \simeq (13, 10)$$

of Figure 4.8, so that the two flows regenerating the love of Roxane are

$$\gamma_1 A_2^* = 0.25 \qquad R_1^L(x_2''') \simeq 1.$$

As the reaction to love is definitely greater than the reaction to appeal, the model is in agreement with Roxane's statement.

4.4 Conclusions

We have confirmed, through the study of the love story described by Edmond Rostand in "Cyrano de Bergerac," the idea discussed in the previous chapter, namely, that temporary bluffing can be rewarding in love affairs. This discovery, consistent with observed behaviors in real life, allows temporary bluffing to be considered as a sort of therapy that could be suggested to couples who would not otherwise be able to achieve a satisfactory romantic relationship. In the context of love affairs, this, in a sense, lessens the negative moral value usually attached to bluffing in the context of social behavior.

Since "Cyrano de Bergerac" stresses the dichotomy between mind and body, its successful interpretation through a mathematical model provides hope that it may be possible to use formal tools to interpret other important works where the same dichotomy is discussed. Along the same lines, we might even hope to be able to give answers to extremely complex dilemmas, like that described by Thomas Mann in his novel "The Transposed Heads". In that novel Sita is married to the intellectual Shridaman, but attracted to the earthy Nanda, a good friend of her husband. The two friends behead themselves. Then, one day, their heads are magically restored—but to the wrong bodies. From that point on, Sita is confused. She does not know which one her husband is and which man she prefers. She is thus forced to face an intriguing dilemma: is it the mind or the body that creates and rules the person? Although Sita's story is undoubtedly more complex than that of Roxane and Cyrano, the use of a rational approach based on a mathematical model could possibly cast light on the story and help in solving the dilemma it poses.

Chapter 5

The discovery of hidden components of the appeal

This chapter is also devoted to the study of standard couples, even if our message, namely, that "small discoveries can have great consequences in love affairs," holds true for many other kinds of couples. From a formal point of view, these great consequences are the result of so-called catastrophic bifurcations. They are important because in general they are associated with great emotions that emerge when there are dramatic breakdowns or explosions of interest. Examples of the first kind are all relationships based mainly on sex. Indeed, sexual appetite inexorably decreases over time so that a point can easily be reached where separation is both unavoidable and unexpected if the individuals are mainly interested in sex. A well known example of the second type is playboys who systematically reinforce their appeal until their prey falls in love with them.

After a short discussion of catastrophic bifurcations in the model, we present the detailed study of the love story between "Beauty and The Beast", a traditional fairy tale released in 1991 as an animation by Walt Disney, and between Elizabeth and Darcy described in "Pride and Prejudice" by the English novelist, Jane Austen. The reason for limiting our attention to the model described in Chapter 3 is that a model like this is perfectly suited for representing the characters of the individuals involved in the two love stories.

No new mathematical prerequisites are needed for reading this chapter. Extra details can be found in Rinaldi *et al.* (2014, 2013b).

5.1 Catastrophic bifurcations

As already shown, standard couples (see model (3.2) in Chapter 3) have two alternative stable states if their appeals belong to the compact white region in Figure 5.1a, from now on called ASS (Alternative Stable States). In contrast, when the point (A_1, A_2) does not fall in ASS only one stable equilibrium exists and the signs of its components are as indicated in Figure 5.1a. Obviously, the signs of the feelings can also be specified in ASS if reference is made to only one of the two alternative stable states. In particular, if reference is made to the equilibrium reached when the two individuals are initially indifferent to each other, the signs of the feelings in the long term are as indicated in Figure 5.1b.

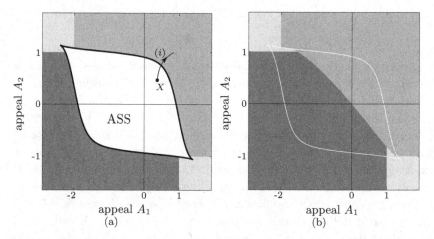

Fig. 5.1   The signs of the feelings in model (3.2) of Chapter 3 for all possible values of the appeals A_1 and A_2. In the green [red] regions the feelings are both positive [negative], while in the yellow regions they are of opposite sign. Parameter values are $\alpha_1 = 0.1$, $\alpha_2 = 0.3$, $\gamma_1 = \gamma_2 = 1$, and reaction functions are as in Figure 3.1 of Chapter 3. (a) The signs of the feelings in the long term. In the white ASS area the signs cannot be specified because in that region there are two alternative stable equilibria. (b) The signs of the feelings in the long term when the two individuals are initially indifferent to each other.

The points of the boundary of ASS are *saddle-node* bifurcation points (see Appendix) and there are no other catastrophic bifurcations in the system. More precisely, for (A_1, A_2) in ASS, there are three equilibria x', x'', and x''' that are ordered from low to high in the sense that $x' < x'' < x'''$. The intermediate equilibrium x'' is always unstable (a saddle), while the two others are stable. On the lower boundary of ASS $x'' = x'''$, while on the upper boundary of the same region $x' = x''$. Thus, consider a couple represented by a point $X = (A_1, A_2)$ in the ASS area close to its upper boundary and assume that the couple is in the unsatisfactory stable state x'. Then, if the appeals slowly increase along path (i) in Figure 5.1a, after some time the point (A_1, A_2) will be just above the upper boundary of ASS where the satisfactory state x''' is the unique equilibrium point. This means that a very small variation in appeal can trigger a remarkable variation in feelings—the transition from state x' to state x'''.

Many real love stories reported in the technical literature, as well as many fictitious stories portrayed in novels and films, show that catastrophic transitions are frequent. This is because the appeals can vary slowly over time due to aging or adaptation, or because a series of small discoveries can force one of the two individuals to gradually change his/her perception of the appeal of the partner. This is exactly the case described in the next two sections, where two famous love stories are analyzed.

5.2   Beauty and The Beast

"Beauty and The Beast" (BB) is one of the most famous European fairy tales (the first trace can be found in the "*Metamorfosi*" by Lucio Apuleio (125-175 AD)). The most popular version, written in French by Jeanne-Marie Leprince de Beaumont in 1756, has been made into films (the first directed by Jean Cocteau in 1946), cartoons, and musicals. Here we refer to the 1991 Walt Disney production which was nominated for best film and won two Oscars (soundtrack and original song).

Twelve short segments of the film (BB1,...,BB12), available at

```
home.deib.polimi.it/rinaldi/BeautyAndTheBeast
```

are described below to support the mathematical interpretation of the love story. Still frames of the twelve film segments are shown in the panels of Figure 5.2 where the initial and final times of each segment are reported in the corresponding panel.

The film begins (BB1) with a short foreword: as a punishment a young prince is transformed by a fairy into a physically repellent beast. He will recover his human aspect only if a young lady explicitly declares herself to be in love with him. Beauty, a very happy and pretty girl (BB2), obtains permission from The Beast to take the place of her father, who is in prison in the tower of The Beast's castle, as he is too old to endure his sentence (BB3). The Beast's friends suggest that Beauty provides a unique opportunity for him to regain his original looks (BB4). Thus, The Beast starts, in his own way, to woo her. First, he transfers her from the prison to a beautiful room, and then invites her to dinner (BB5) in an unrefined man-

Fig. 5.2   Still frames of the twelve film segments (BB1,...,BB12) described in the text.

ner. Beauty refuses the invitation with disdain (BB6). Some time later, Beauty, attacked in the woods by a horde of wolves, is rescued by The Beast who shows with great courage that he is ready to sacrifice himself for her (BB7). Taking care of his wounds, Beauty thanks him, demonstrating for the first time a slightly less hostile attitude toward him (BB8). Encouraged by this sign, The Beast offers her access to the beautiful castle library, a gift that she definitely appreciates (BB9). It is after this gesture that Beauty perceives the altruism of The Beast, and a little later, in an idyllic scene, she shows the first signs of emotional involvement (BB10). The inevitable transition to a state of reciprocated love has already started and culminates with the dance in the great hall of the castle during which she rests her head graciously on his chest and admits to being happy (BB11). At this point the story finishes, with Beauty happy in spite of the still repellent physical aspect of The Beast. To crown the fairy tale the last scene (BB12) shows The Beast dying in the arms of Beauty, who declares her love to him, thus triggering the magic of his transformation into a prince.

The love story just described can be perfectly interpreted using model (3.2) of Chapter 3 provided suitable values are assigned to the appeals of the two individuals. Beauty's appeal A_1 is definitely high (BB2) and remains so during the whole story. The appeal of The Beast (perceived by Beauty) A_2 has a negative component due to his repellent physical aspect and a positive component due to his altruism. The negative component is large and fixed, while the positive component increases over time because it is gradually discovered by Beauty through a series of specific events (BB5,7,9) (in reality, in the film there are also other episodes—not mentioned here for the sake of brevity—during which Beauty has the chance of appreciating the altruism of The Beast). The evolution of the love story can thus be seen as resulting from a slow variation in the appeals along the path indicated in Figure 5.3a. At the beginning of the story (point 1 in Figure 5.3a)— soon after Beauty and The Beast have met—her feeling at equilibrium is negative while his feeling is positive (yellow region in Figure 5.3a). Then, when A_2 increases, nothing relevant occurs (in the sense that the signs of the feelings remain unchanged) until the upper saddle-node bifurcation curve is crossed from below (BB9). This crossing triggers a relevant transition in the space of the feelings that ends at a positive equilibrium (point 4 is in the green region in Figure 5.3a) even if the appeal of The Beast is still negative (BB11).

The analysis of the null-clines reported in Figure 5.3b for various values of A_2 and the graph of Beauty's love versus The Beast's appeal reported in Figure 5.3c are also interesting because they show that the feeling of Beauty increases progressively but remains negative until the catastrophic transition toward the positive equilibrium (point 4 in Figure 5.3b) occurs.

It is interesting to note that in the film the description of the first symptoms of the catastrophic transition (see BB10) is accompanied by a beautiful melody in which the friends of The Beast repeatedly mention (via the voice of Celine Dion) the discontinuity associated with the catastrophic transition:

> *There may be something there that wasn't there before*
> *You know, perhaps there's something there that wasn't there before.*

Obviously, this is not a pure coincidence but reveals a great deal of skill on the part of those who wrote this hymn to catastrophic transitions, justifying, one might say, the Oscar award for the best original song.

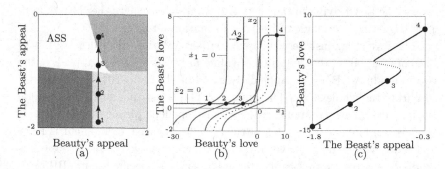

Fig. 5.3 Interpretation of the love story between Beauty and The Beast with model (3.2) of Chapter 3. Points 1, ..., 4 indicate the different values of the appeals during the story and the corresponding values of the feelings at equilibrium. (a) The Beast's appeal (perceived by Beauty) varies during the story because she becomes gradually aware of his altruism through a series of small discoveries (point 2: BB5; point 3: BB7; point 4: BB9). (b) The null-cline $\dot{x}_1 = 0$ shifts to the right when A_2 is increased, while the null-cline $\dot{x}_2 = 0$ remains unchanged. The dotted null-cline corresponds to the saddle-node bifurcation. (c) Beauty's love at equilibrium as a function of The Beast's appeal. The dotted segment of the curve corresponds to saddle equilibria.

5.3   Elizabeth and Darcy

"Pride and Prejudice" (Austen, 1813) is a very popular English novel in which the love story between Elizabeth and Darcy is described. The novel has inspired a number of films, the most successful of which was released in 2005, starring Keira Knightley as Elizabeth and Matthew Macfadyen as Darcy. Fifteen short segments of the film (PP1,...,PP15), available at

> `home.deib.polimi.it/rinaldi/PrideAndPrejudice`

are described below to support the mathematical interpretation of the love story. Still frames of the fifteen film segments are shown in Figure 5.4, where the initial and final times of each segment are reported on the corresponding panel.

Fig. 5.4 Still frames of the fifteen film segments (PP1–PP15) described in the text.

Elizabeth, the 20-year-old second-born of a modest family, lives with her parents and sisters in a lovely country house in Hertfordshire (PP1). At a ball she is introduced to Darcy, an elegant young man with an arrogant air who belongs to the wealthy local nobility (PP2). Unbeknownst to Darcy, Elizabeth hears him making a disparaging remark about her appeal (PP3). Her reaction is strong and bitter (PP4). This is only the first in a series of disputes in which she displays ideas that are strongly in conflict with those of Darcy (PP5) or members of his family (PP7). Sometimes these disputes are as intense as real conflicts, such as when they dance together for the first time (PP6). However, Darcy is fascinated by her grace and talents (PP8) to the point of showing the first symptoms of involvement when they are first alone for a short time (PP9). Soon after that, Darcy declares his love, mentioning, however, the difficulties that the difference in their social status implies for a marriage (PP10). Elizabeth does not appreciate his arguments and, when he proposes, refuses to marry him (PP10). Upset by her refusal, Darcy writes a letter to Elizabeth in which he apologizes, at the same time justifying in his detailed and convincing arguments the very points of his past behavior that she disliked in him (PP11). This letter is the turning-point of the entire story because it removes the obstacles that prevented her from fully appreciating his excellent qualities. Her involvement grows inexorably from a state of initial confusion (PP12) to acknowledging his appeal (PP13) and from jealousy (PP14), when she suspects he is in love with someone else, to happiness (PP15), when she discovers that her jealousy is unjustified. Thus, the awareness of a fully requited love is gradually reached and culminates with a romantic kiss in the last scene of the film.

The love story between Elizabeth and Darcy can be perfectly interpreted using model (3.2) of Chapter 3, provided suitable values are assigned to the parameters of the model. The appeal A_1 of Elizabeth (perceived by Darcy) is initially low (PP3) but then increases at each encounter, where she shows, with no exception, her grace and talent (PP5–PP8). In contrast, the appeal A_2 of Darcy (perceived by Elizabeth) remains negative, in view of the prejudices she has against the rich and the upper classes. The evolution of the love story can thus be seen as the result of a recursive increase of A_1 followed by a final sudden increase of A_2 (due to Darcy's revealing letter) as shown in Figure 5.5a. At the beginning of the story (PP2,3,4) the representative point in the space of the appeals is point 1 in the red region of Figure 5.5a where Elizabeth and Darcy are in an antagonistic relationship. When the perception A_1 of Elizabeth's appeal increases (see points

2,3,4,5 in Figure 5.5a) nothing relevant occurs though Darcy's involvement is positive at point 5 (see PP9). At that point Elizabeth is still antagonistic and, indeed, she refuses to marry him (PP10). It is only Darcy's letter that suddenly reveals his honor to Elizabeth. This is represented as a vertical jump in the space of the appeals from point 5 to point 6 in the green region of Figure 5.5a where the two can only be in a positive relationship. In other words, as a consequence of the letter the upper saddle-node bifurcation curve is crossed from below and this crossing implies a discontinuous jump from x' to x''' in the feelings of Elizabeth and Darcy (PP12–PP15).

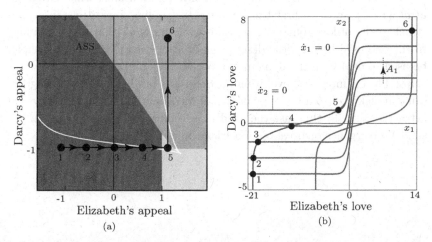

Fig. 5.5 Interpretation of the love story between Elizabeth and Darcy. Points 1,...,6 indicate the different values of the appeals during the story and the corresponding values of the feelings at equilibrium. (a) Elizabeth's appeal (perceived by Darcy) increases because he becomes gradually aware of her grace and talents through a series of encounters (point 2: PP4; point 3: PP6; point 4: PP7; point 5: PP8). Darcy's appeal (perceived by Elizabeth) suddenly increases when he writes a letter to her (PP11). (b) The null-cline $\dot{x}_2 = 0$ shifts upward when A_1 increases, while the null-cline $\dot{x}_1 = 0$ jumps to the right when she receives the letter from him.

The analysis of the null-clines in the love space reported in Figure 5.5b is also interesting. It shows that the blue null-cline shifts upward when A_1 increases (from point 1 to point 5 in Figure 5.5a). This implies that the involvement of Darcy increases gradually and finally becomes positive at point 5, where, however, Elizabeth's involvement is still negative. But Darcy's letter suddenly shifts the red null-cline so much to the right that the two null-clines intersect only at a single point, namely, point 6 in the positive quadrant.

5.4   Conclusions

The analysis in this chapter points out a general property—known from personal experiences—namely, that a series of small discoveries can give rise to a sudden turning-point in the development of a love story. In mathematical terms these turning-points are so-called catastrophes, which, in the cases of "Beauty and The Beast" and of "Pride and Prejudice," are technically revealed by the existence of a saddle-node bifurcation. As far as we know, Jones (1995) was the first to invoke, on purely intuitive grounds, the use of catastrophe theory in the study of romantic relationships, while Gragnani *et al.* (1997) were the first to discover catastrophic bifurcations in an abstract model of love dynamics. It is also worth noting that models similar to the one used here for the description of a love story could, in principle, be used to deal with the dynamics of other kinds of interpersonal relationships. For example, the evolution of friendship can be described using the same model, provided the appeals do not have anything to do with physical attractiveness. In this case the saddle-node bifurcation would explain the sudden emergence or breakup of a friendship.

Chapter 6

Couples composed of insecure and unbiased individuals

In this chapter we study couples composed of insecure and unbiased individuals. Thus, the only difference with respect to the case of standard couples considered in Chapters 3, 4, and 5 is that the reaction functions first increase and then decrease with the love of the partner. This implies that the geometry of the two null-clines is different from that in the previous chapters, but that variations of the appeals still shift (vertically or horizontally) the two null-clines. For this reason, the main properties of the couple can still be derived from simple geometric considerations based on the geometry of the null-clines. For simplicity, we study only the case in which the reaction to love increases up to a threshold value of the love of the partner and then declines but remains positive. This implies that the analysis can be performed by looking only at trajectories in the positive quadrant of the space of the feelings because initially non-antagonistic partners remain so forever. Moreover, the reaction to love is described by a function identified by two parameters interpretable as reactiveness to love and degree of insecurity (higher insecurities characterize reactions to love that start declining earlier).

Our analysis starts with the case of individuals differing only in their appeals or in their insecurity and then proceeds with the study of individuals that also differ in another or many other traits. The result of the analysis is that couples composed of insecure and unbiased individuals have either a single (globally stable) equilibrium or two alternative stable equilibria, exactly as in the case of standard couples. Moreover, one of the two alternative stable states is more rewarding for "she" and the other for "he", while in the case of standard couples one of the two alternative stable states was more rewarding than the other for both partners. This is clearly indicated by our figures where in the pink [blue] regions she [he] is more involved than he [she] if the love story starts from the state of indifference. Since the unbalance in her and his feelings is often quite consistent, our analysis predicts that in these couples one of the two individuals, actually the more insecure and/or appealing one, is definitely less satisfied than the partner, and is therefore inclined to break off the relationship. This means that in the pink [blue] region she [he] has high chances of being abandoned by the partner. This property, here derived from purely theoretical arguments, is consistent with results obtained from empirical studies.

No new mathematical prerequisites are needed for reading this chapter.

6.1   The model

The couples we consider in this chapter are composed of insecure individuals, who, by definition, are characterized by their reactions to love first increasing and then decreasing with the love of the partner. Two typical reaction functions are shown by curves (iii) and (iv) in Figure 1.6 of Chapter 1. Both curves describe individuals who react more and more positively to the love of the partner at low involvements and less and less strongly at high involvements, that is, above the threshold corresponding to the maximum of the curve. The difference between the two curves is that in case (iii) the reaction, though declining, remains positive, while in case (iv) it becomes negative. Thus, an individual of type (iv) is so annoyed by the high involvement of the partner that his/her reaction to love—usually a regenerating flow—becomes a consumption flow (like that associated with oblivion).

Here and in the next chapter, we consider only reaction functions of type (iii) and we actually limit our analysis to the following particular class of functions characterized by two parameters (β_i, k_i)

$$R_i^L(x_j) = \beta_i x_j \exp(-k_i x_j). \tag{6.1}$$

The graph of the function (6.1), represented in Figure 6.1 for $\beta_i = 15$ and $k_i = 1$, has a maximum R_i^* for $x_j = x_j^*$.

Since

$$\frac{dR_i^L(x_j)}{dx_j} = \beta_i(1 - k_i x_j) \exp(-k_i x_j),$$

the maximum reaction (at which $dR_i^L(x_j)/dx_j = 0$) is obtained for

$$x_j^* = \frac{1}{k_i},$$

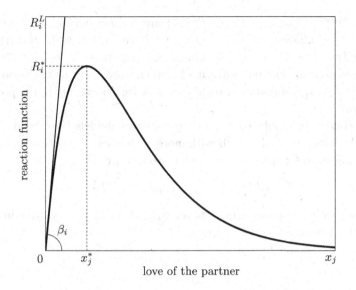

Fig. 6.1   The reaction function $R_i^L(x_j)$ given by equation (6.1).

and the value of the peak R_i^* is

$$R_i^* = R_i^L(x_j^*) = \frac{1}{e}\frac{\beta_i}{k_i},$$

where $e = \exp(1)$. At low values of x_j the reaction R_i^L can be approximated as

$$R_i^L(x_j) \sim \frac{dR_i^L(x_j)}{dx_j}\bigg|_{x_j=0} x_j = \beta_i x_j.$$

Thus, at low involvements the reaction to love is the same as in linear models (see Chapter 2).

In conclusion, the parameter β_i is the reactiveness to love (at low involvements of the partner), while the parameter k_i controls the partner's love x_j^* at which the reaction function peaks. In particular, individuals with high values of k_i have a low x_j^*, that is, they are annoyed by the partner's love earlier than individuals with low k_i. This means that k_i is an indicator of the insecurity of the individual.

A couple composed of two insecure individuals is therefore identified by 10 positive parameters, namely $(\alpha_i, \beta_i, \gamma_i, k_i, A_i)$, $i = 1, 2$, and the equations of the model are

$$\begin{aligned}
\dot{x}_1 &= -\alpha_1 x_1 + R_1^L(x_2) + \gamma_1 A_2 \\
\dot{x}_2 &= -\alpha_2 x_2 + R_2^L(x_1) + \gamma_2 A_1,
\end{aligned} \tag{6.2}$$

where the two return functions $R_i^L(x_j)$ are as in equation (6.1). On the positive x_1-semiaxis ($x_1 \geq 0$, $x_2 = 0$) we have $\dot{x}_2 \geq 0$ and, similarly, on the positive x_2-semiaxis $\dot{x}_1 \geq 0$. This means that trajectories can enter the positive quadrant but not leave it. In other words, the model is positive and this is why, from now on, state portraits are drawn only in the positive quadrant.

Of course, we could consider more complex models by extending the analysis to insecure individuals with more sophisticated reaction functions. For example, in Chapter 8 we consider reaction functions

$$R_i^L(x_j) = \beta_i k_i x_j \exp[-(k_i x_j)^{n_i}]$$

identified by three parameters (β_i, k_i, n_i), where n_i is a positive integer. Since this reaction function peaks at

$$x_j^* = \frac{1}{k_i} \sqrt[n_i]{\frac{1}{n_i}},$$

the parameter k_i is still a measure of the insecurity of individual i.

6.2   The case of identical individuals

To determine the complete catalog of possible behaviors of model (6.2) we start with the analysis of the very special case of identical individuals ($\alpha_1 = \alpha_2 = \alpha$, $\beta_1 = \beta_2 = \beta$, $\gamma_1 = \gamma_2 = \gamma$, $k_1 = k_2 = k$, $A_1 = A_2 = A$).

The two null-clines $\dot{x}_1 = 0$ and $\dot{x}_2 = 0$, given by

$$x_1 = \frac{\beta}{\alpha} x_2 \exp(-kx_2) + \frac{\gamma}{\alpha} A$$
$$x_2 = \frac{\beta}{\alpha} x_1 \exp(-kx_1) + \frac{\gamma}{\alpha} A,$$

are, therefore, symmetric with respect to the straight line $x_2 = x_1$, as shown in Figure 6.2 for the parameter settings specified in the caption. The intersection of the two null-clines (*i.e.*, the equilibria of the system) are either three (one with $x_1 = x_2$), as in Figure 6.2a, or only one (with $x_1 = x_2$), as in Figure 6.2b. The directions of the trajectories on the null-clines suggest the nature of the equilibria. In Figure 6.2a, the equilibrium with $x_1 = x_2$ is a saddle, while the two others are stable nodes. In contrast, in Figure 6.2b the equilibrium is a stable node. Of course, one can verify these results analytically by determining the eigenvalues of the Jacobian matrix evaluated at each equilibrium.

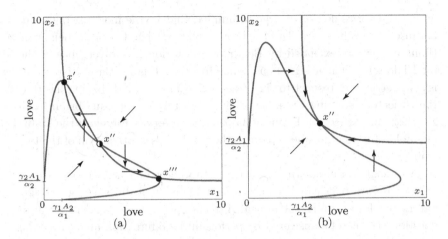

Fig. 6.2   The null-clines of model (6.2) with identical individuals ($\alpha_1 = \alpha_2 = \alpha = 1$, $\beta_1 = \beta_2 = \beta = 15$, $\gamma_1 = \gamma_2 = \gamma = 1$, $k_1 = k_2 = k = 1$, $A_1 = A_2 = A$). (a) $A = 1$: three equilibria (x' and x''' stable nodes, x'' saddle). (b) $A = 3$: one equilibrium (x'' stable node).

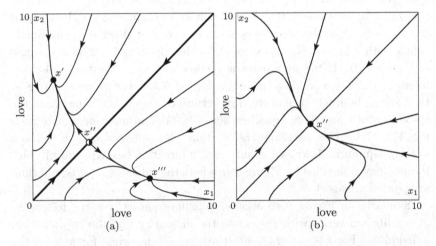

Fig. 6.3   State portraits for the parameter values specified in the caption of Figure 6.2.

The state portraits corresponding to Figures 6.2a and 6.2b are shown in Figures 6.3a and 6.3b. They can be produced by simulation for any parameter setting, but their main properties can be obtained through simple and general theoretical arguments. As already mentioned, if the feelings are initially positive they remain positive forever (positivity of the model). Moreover, trajectories do not diverge to infinity, that is, the model is bounded, because large values of x_1 [x_2] imply $\dot{x}_1 < 0$ [$\dot{x}_2 < 0$] and periodic behaviors (limit cycles) are excluded by Bendixon's criterion (the divergence of the model does not change sign, being equal to -2α). Finally, the state portrait is symmetric with respect to the straight line $x_1 = x_2$ and this implies that the trajectory starting from the origin tends toward the equilibrium with $x_1 = x_2$. In the case of Figure 6.3a, this trajectory is therefore the stable manifold of the saddle and separates the two basins of attraction of the two stable nodes.

A crucial and interesting question is to understand how the state portrait of Figure 6.3a can become that of Figure 6.3b by varying a single parameter, for example, the appeal A in the specific case. This is a typical question that can be answered by performing a bifurcation analysis of the system. In the present case this can be done without any computation being needed, by simply looking at the geometry of the null-clines. In fact, an increase in the appeal A has the effect of shifting the first null-cline ($\dot{x}_1 = 0$) to the right, and the second ($\dot{x}_2 = 0$) upward by the same amount. Thus, if we start from the state portrait of Figure 6.3a (where $A = 1$) and we continuously increase A, the central equilibrium moves along the straight line $\dot{x}_1 = x_2$, while the two others become closer and closer to it and finally collide with it for a particular value A^* of the appeal ($A^* = 2.33$ in the case of Figure 6.3). If the parameter is further increased, the two null-clines intersect only at a single point as in Figure 6.3b. For $A = A^*$ there is, therefore, a bifurcation, namely, a structural change in the behavior of the system (which has three equilibria for $A < A^*$ and only one equilibrium for $A > A^*$). This particular bifurcation, characterized by the collision of three equilibria, is known as *pitchfork* bifurcation (see Appendix). This terminology is justified by Figure 6.4a where the values of x_1 at equilibrium are plotted versus A.

Similarly, in Figure 6.4b another pitchfork shows how the feeling x_1 at equilibrium varies with respect to the insecurity k of the two identical individuals. For $k < k^*$ there are three equilibria, while for $k > k^*$ the equilibrium is unique. Figure 6.4b can also be predicted by looking at the null-clines. In fact, if k is increased, the peak (x^*, R^*) of the return function (see Figure 6.1) moves to the left (because $x^* = 1/k$) and is attenuated (because $R^* = \beta/(ek)$), so that the three equilibria in Figure 6.2a get closer and closer until they collide for a particular value of k, which we have called k^*. Then, for $k > k^*$ the null-clines are as in Figure 6.2b. In other words, if we vary the insecurity k we observe the same phenomena

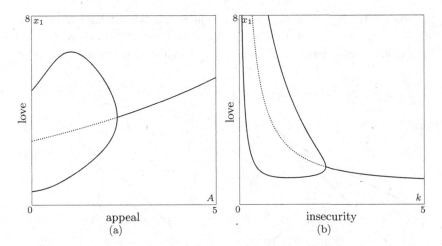

Fig. 6.4 Typical pitchfork diagrams of model (6.2) with identical individuals. Parameters α, β, γ as in the caption of Figure 6.2. (a) $k = 1$. (b) $A = 1$.

that we discovered by varying the appeal A. Using similar arguments the reader can verify that the same bifurcations are also obtained by varying the other parameters, namely, α, β, and γ.

6.3 The general case

Although the aim of this section is to analyze generic couples of insecure and unbiased individuals, we first consider the case of individuals that differ only in one trait, say, the appeal. If the appeals A_1 and A_2 of the two individuals coincided, that is, $A_1 = A_2 = A$, then we would be in the case analyzed in the previous section using either $A < A^*$ or $A > A^*$. If we assume the first case, the null-clines are as in Figure 6.2a and the effect of a perturbation of one of the appeals can be easily detected by shifting only one of the two null-clines. For example, if we gradually increase A_1, then the null-cline $\dot{x}_2 = 0$ moves upward and the equilibria x'' and x''' get closer and closer until they finally collide and disappear. As we have already seen in the preceding chapters, this is a saddle-node bifurcation. But the same occurs if we gradually decrease A_1 because in this case x' and x'' get closer and closer and finally collide and disappear. Moreover, the symmetry of the model implies that the same saddle-node bifurcations are obtained by varying A_2.

As what we have said holds true for any $A < A^*$ we can infer that in the space (A_1, A_2) of the appeals two saddle-node bifurcation curves must merge at a point (A^*, A^*). This is shown in Figure 6.5a, where the two bifurcation curves have been obtained through continuation. The point

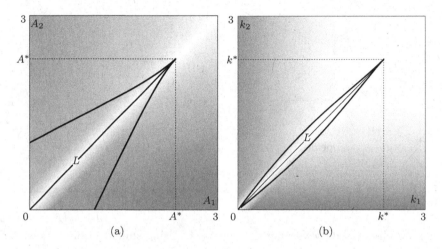

(a)   (b)

Fig. 6.5   Region (cusp) where couples with individuals differing in a single trait (appeal in (a) and insecurity in (b)) have two alternative stable states. Curve L separates couples converging to x' from couples converging to x''' when starting from the state of indifference. The colors indicate the points at which she (pink) or he (blue) are more involved at the end of the story. Parameter values are $\alpha_1 = \alpha_2 = 1$, $\beta_1 = \beta_2 = 15$, $\gamma_1 = \gamma_2 = 1$. (a) $k_1 = k_2 = 1$. (b) $A_1 = A_2 = 1$.

(A^*, A^*) is called *cusp*, as is the region delimited by the two curves. For any point (A_1, A_2) in the cusp region the model has two alternative stable states x' and x''' and one saddle x''. The cusp region can be partitioned as shown in Figure 6.5a by a curve L which is the locus of pairs (A_1, A_2) for which the trajectory starting from the origin tends toward the saddle x''. The symmetry of the model implies that the curve L is the straight line $A_1 = A_2$. Moreover, the romantic regime asymptotically reached by individuals initially indifferent to each other is x' if point (A_1, A_2) is below curve L and x''' if it is above it. That is to say, if she is more appealing than he is, the relationship evolves toward a romantic regime in which he is more satisfied than she is, because $x'_2 > x'_1$. And, obviously, the opposite holds true if he is more appealing than she is. All this is clearly indicated by the colors in Figure 6.5.

Analogous results are obtained if the individuals differ in another trait. For example, Figure 6.5b describes the case of individuals differing only in their insecurity. The conclusion is that the individual who is more insecure is the one who is less involved at the end of the story.

In the general case of individuals with different traits, the symmetries that we have exploited in our previous analysis break down. But the results remain qualitatively the same because the geometric arguments we have used to discuss the bifurcations with the null-clines are still valid. Thus, the saddle-node bifurcation curves delimiting the cusp region where there are alternative stable states and the curve L partitioning the cusp can only be obtained through numerical bifurcation analysis and continuation. Figure 6.6 shows two examples of the typical results one should expect for couples of generic individuals. The pink and blue colors in the figure point out the regions where she or he are more involved at the end of the story (starting from the state of indifference). Again, the final message is that the final involvements are unbalanced in favor of the less appealing and less insecure individual.

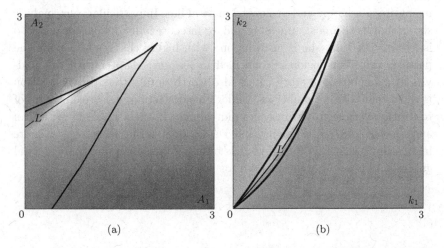

Fig. 6.6   Region (cusp) where couples with individuals with different traits have two alternative stable states. Curve L and colors as is Figure 6.5. Parameter values are $\alpha_1 = 0.9$, $\alpha_2 = 1.1$, $\beta_1 = 14$, $\beta_2 = 16$, $\gamma_1 = \gamma_2 = 1$. (a) $k_1 = 1.1$, $k_2 = 0.9$. (b) $A_1 = 2$, $A_2 = 0.5$.

6.4   Conclusions

The results we have obtained in this chapter for insecure and unbiased individuals can usefully be compared with those derived for secure and unbiased individuals in Chapter 3. Couples of secure and unbiased individuals can have one or two romantic regimes but the existence of one (and only one) regime satisfactory for both individuals is always guaranteed. This gives these couples a high chance of remaining for a long time (if not forever) in a high quality romantic regime as established in Tracy *et al.* (2003) through the analysis of questionnaires.

Conversely, insecure individuals can have two romantic regimes, both characterized by positive involvements of the partners. However, our analysis has shown that if the partners have diversified but constant traits, in particular appeals and/or insecurities, the romantic regimes are very unbalanced. In the real life of a couple such a feature introduces a relevant source of friction that can easily end in the breakdown of the relationship unless some of the traits vary over time in a more or less random way that favors each of the two partners alternately (Banerjee *et al.*, 2015) thus maintaining a happy and prolonged marital life. In particular, our analysis shows that as the most appealing and/or insecure individual is the one who is less involved, he/she is also the one with higher chances of leaving the partner. This conclusion is confirmed in the next chapter with the analysis of a famous love story. It is also worth noting that this conclusion is consistent with a number of studies on satisfaction and breakup rate of romantic relationships of insecure individuals (Hatfield *et al.*, 1989; Hazan and Shaver, 1987; Kirkpatrick and Davis, 1994). But what is even more interesting is that this conclusion has been obtained on a purely theoretical grounds, that is, without the need for any empirical evidence.

Chapter 7

Scarlett and Rhett

This chapter is devoted to the study of the love story described in "Gone with the Wind", one of the most popular films of all times. The film, released in Atlanta in December 1939, starring Vivien Leigh as Scarlett and Clark Gable as Rhett (see Figure 7.1), was inspired by Margaret Mitchell's 1936 bestseller. It was one of the first films in color, was awarded eight Oscars, has been seen at least once by 90% of Americans, and was, financially, the most successful film ever, until a few years ago.

Vivien Leigh   Clark Gable

Scarlett O'Hara   Rhett Butler

Fig. 7.1 Shots of the film "Gone with the Wind".

The love story between Scarlett and Rhett is a perfect example of the theory developed in the previous chapter. Both lovers are undoubtedly insecure and their love story can be split into two parts separated by the Civil War. In the first part, she is more appealing than he is and, as in the theory expounded in the previous chapter, after he declares his love she refuses him. In contrast, after the war is over, his appeal is definitely higher as he has become very rich, while she, having financial problems, is more interested in money. Thus, their love story develops in a completely different way and ends with Scarlett desperately asking him to stay and Rhett leaving the house with the famous quote "Frankly, my dear, I don't give a damn". It is also worth noting that the agreement between the story predicted by the model and the story described in the film can be supported by some qualitative but interesting arguments.

No new mathematical prerequisites are needed for reading this chapter. Some extra details can be found in Rinaldi *et al.* (2013a).

7.1   The love story between Scarlett and Rhett

The love story between Scarlett and Rhett develops during the decade starting with the Civil War in 1861. In comparison with the film, the story is here extremely compressed, eliminating a number of details concerning minor characters and historically important events. Figures 7.2 and 7.3 report the initial and final times of twelve short segments (S1–S12) of the film available at

    home.deib.polimi.it/rinaldi/GoneWithTheWind

and used for deriving and supporting the model.

The film starts with a barbecue party where the beautiful Scarlett flirts with a circle of admiring men (S1). She responds promptly and positively even to small signs of interest on the part of her suitors, but then quickly changes her mood as soon as she has obtained a minimal success. This is the typical behavior of a conqueror, first exciting and then rebuffing her victim. This easily identifiable tendency on the part of Scarlett to avoid too heavy emotional involvements is possibly due to her childish infatuation for Ashley. To sum up, her reaction is positive for low involvement on the part of any potential partner but then declines sharply when the pressure exerted on her increases so much that her dream with Ashley is called into question.

At the same party, Scarlett and Rhett see each other for the first time (S2). He is tall, attractive, very elegant, and looks like a professional "Don Juan". This impression is also generated by the quite impertinent way he looks at her, and reinforced by the news that he has seduced a Charleston girl and then refused to marry her.

(S1) Scarlett flirting at a party

(S2) The first encounter

(S3) Rhett teases Scarlett
about her widowhood

(S4) Rhett bids to dance with
Scarlett and she accepts

(S5) The first kiss

(S6) Rhett declares his love
before leaving her

Fig. 7.2   Still frames of the six film segments (S1–S6) used in the text to describe the first part of the story.

One year later Scarlett and Rhett meet again just after she has lost her first husband (S3). They meet at a dance party organized to raise money

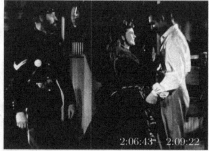

(S7) Scarlett visits Rhett in jail

(S8) Rhett proposes marriage to Scarlett

(S9) Scarlett and Rhett
just after marrying

(S10) Strong signs of
Rhett's disinvolvement

(S11) Rhett offers Scarlett
a divorce

(S12) Rhett leaves Scarlett

Fig. 7.3　Still frames of the six film segments (S7–S12) used in the text to describe the second part of the story.

for the Confederate Cause, and Rhett teases Scarlett about her widowhood, revealing that he knows that she is there because of her devotion not to the

Cause but to enjoyment.

One of the organizers causes a stir by announcing that to raise funds he is auctioning the right to dance with the ladies (S4). Rhett bids a hundred and fifty dollars in gold for Scarlett, thus showing a definite interest in her. The organizer protests that Scarlett cannot dance because she is in mourning and asks him to change his bid to another lady, but Rhett refuses. To everyone's shock, Scarlett accepts and rushes onto the dance floor.

Rhett grows impatient as Scarlett continues to wear black mourning clothes even though she is taking part in all social activities (S5). To tempt her to discard her black veil, he buys her a fashionable bonnet in Paris. Scarlett cannot resist the bonnet but she warns him that she will not marry him in return. He replies that he is not the marrying kind—though he kisses her, thus revealing once more his involvement.

In August 1864 when Atlanta is burning, Rhett rescues Scarlett from the Yankees and tells her that he is going to join the Confederate army (S6). He says, in a light-hearted tone, that in spite of what he said to her previously, he does love her, because they are alike. He kisses her passionately, and she is surprised to feel herself responding. But she is still furious at him for deserting her, so she slaps his face and declares that he is not a gentleman. This is the end of the first part of the story.

In 1865 when the war is over, Scarlett is beset by financial problems related to cotton production at her Tara plantation. She suddenly realizes that Rhett, who has become very rich in the meantime, could solve her problems (S7). Scarlett puts on her new dress and goes to visit Rhett in jail. She pretends to be distressed about his plight, claiming that she would suffer if he were hanged, while in reality she is mostly interested in his money.

Some time later, Rhett proposes marriage to Scarlett (S8). Scarlett replies that she does not love him and does not want to marry again, but Rhett says that she married once for spite and once for money, and has never tried marrying for fun. He kisses her passionately, and she kisses him back. Feeling faint, she agrees to marry him. He asks her why she said yes, and she admits that it was partly because of his money.

After marrying, Rhett treats Scarlett with tenderness and allows her to spend as much money as she likes to restore Tara (S9).

One day, during an argument, Scarlett threatens to stop sleeping with Rhett. He responds with indifference, saying that there are plenty of other women's beds in the world (S10). Scarlett is mortified that he has taken her threat so lightly and is angry at his rudeness.

One night, after a party, Rhett, who has drunk too much, makes violent love to Scarlett. For the first time, Scarlett feels that she has met someone she cannot bully or break. The next morning, she awakes alone, in a state of wild excitement about the previous night (S11). She feels passionate about Rhett, in spite of his cold attitude. Rhett makes light of the night he spent with Scarlett, apologizing for being drunk and offers her a divorce.

In the last scene of the film (S12) Scarlett insistently says that she loves him more than everything else. Rhett replies that he no longer loves her and all he feels for her now is pity. He is going away, maybe to Charleston to try to make peace with his family. Scarlett desperately asks him what she will do if he goes away. Leaving the house, he replies with the famous quote *"Frankly, my dear, I don't give a damn"*.

7.2   The model

From the discussion in the previous chapter, it is clear that Scarlett and Rhett do not belong to the class of secure individuals. Indeed, their reactions first increase with the love of the partner and then decrease. Thus, in agreement with the previous chapter, the model we propose for Scarlett and Rhett is the following

$$
\begin{aligned}
\dot{x}_1(t) &= -\alpha_1 x_1(t) + R_1^L(x_2) + \gamma_1 A_2 \\
\dot{x}_2(t) &= -\alpha_2 x_2(t) + R_2^L(x_1) + \gamma_2 A_1,
\end{aligned}
\tag{7.1}
$$

where $R_i^L(x_j)$ is the simplest analytical form of the reaction of an insecure individual, *i.e.*,

$$
R_i^L(x_j) = \beta_i x_j \exp(-k_i x_j).
$$

As the film makes it difficult to distinguish between the 4 parameters α, β, γ and k identifying Scarlett and Rhett, we have assumed them to be equal, that is, $\alpha_1 = \alpha_2$, $\beta_1 = \beta_2$, $\gamma_1 = \gamma_2$ and $k_1 = k_2$. In contrast, the appeals of Scarlett and Rhett are different. In the first part of the story, when Scarlett is not so much interested in money, the two individuals have a comparable appeal. However, noticing that

$$
\dot{x}_i(0) = \gamma_i A_j
$$

and comparing the reaction of Scarlett and Rhett (S2) when they see each other for the first time (she says: "He looks as if he knows what I look like without my shimmy!"), it is reasonable to imagine that A_1 is marginally

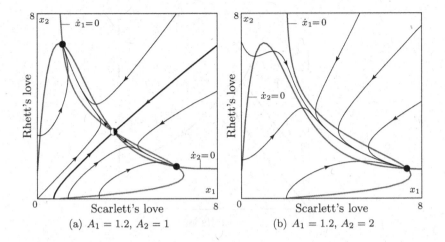

(a) $A_1 = 1.2$, $A_2 = 1$ (b) $A_1 = 1.2$, $A_2 = 2$

Fig. 7.4 Evolution of the love story between Scarlett and Rhett for two different assumptions on their appeals A_1 and A_2. Black curves are trajectories of model (7.1) in love space, red and blue curves are null-clines, half-empty points are saddles and solid points are stable equilibria. In (a) Scarlett is slightly more appealing than Rhett ($A_1 = 1.2$, $A_2 = 1$) and the trajectory starting from the origin tends to the stable equilibrium where Rhett is much more involved than Scarlett. In (b) the appeal of Rhett is definitely greater than that of Scarlett ($A_1 = 1.2$, $A_2 = 2$) and there is only one globally stable equilibrium where Scarlett is much more involved than Rhett.

greater than A_2. For this reason, we have assumed that $A_1 = 1.2$ and $A_2 = 1$.

In the second part of the story, namely, when the war is over and Scarlett has financial problems, the appeal of Rhett is higher, as repeatedly confirmed by Scarlett (S7, S8). For this reason, we have assumed that in this phase $A_1 = 1.2$ and $A_2 = 2$. Under these assumptions, the story should initially evolve along the trajectory starting from the origin in Figure 7.4a and then end, after the war, along a trajectory of Figure 7.4b. The only problem that remains to be solved is the concatenation of the two phases, which is interrupted by a relatively long (6-month) period of separation. Assuming that during the separation period, only the oblivion process is active, we can simply derive the initial conditions of the final phase by integrating the equations

$$\dot{x}_1(t) = -\alpha_1 x_1(t)$$
$$\dot{x}_2(t) = -\alpha_2 x_2(t)$$

starting from the final conditions of the initial phase (since $\alpha_1 = \alpha_2$ the oblivion phase is represented by a straight trajectory pointing toward the

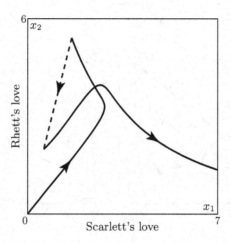

Fig. 7.5 The love story between Scarlett and Rhett predicted by the model. The story starts from the origin when Scarlett and Rhett see each other for the first time. In this phase Scarlett is slightly more appealing than Rhett. Then, a period of separation due to the Civil War follows (dotted trajectory). Finally, a last phase where Rhett is definitely more appealing than Scarlett closes the story.

origin).

The result of this analysis is shown in Figure 7.5, where the dotted trajectory corresponds to the separation phase. The figure shows that in the initial phase Rhett is more involved than Scarlett while after the war the story evolves in the opposite direction with Rhett finally becoming less and less involved. The story predicted by the model tends toward an equilibrium with $x_2 \ll x_1$. As this equilibrium strongly penalizes Rhett, one can easily imagine that a possible consequence in a real context is that Rhett leaves Scarlett when approaching the equilibrium.

7.3 Validation of the model

We now show that the love story predicted by the model (synthetically described in Figure 7.5) is, qualitatively speaking, in good agreement with the story described in the film.

First we place the segments of the film reported in Figure 7.2 along the trajectory as shown in Figure 7.6. This is done by giving a subjective judgment of the involvements of Scarlett and Rhett emerging from each segment. Thus, (S2) is placed at the origin, because $x_1(0) = x_2(0) = 0$

Fig. 7.6   Proposed allocation of the film segments along the trajectory of Figure 7.5. The eleven segments S2–S12 extracted from the film are allocated along the trajectory describing the love story between Scarlett and Rhett following a subjective judgment of their involvements. S2 marks the beginning of the story, S6 and S7 the beginning and the end of the separation period, and S12 the end of the story when Rhett leaves Scarlett.

when Scarlett and Rhett see each other for the first time, while (S6) is placed at the end of the first part of the trajectory because it corresponds to the beginning of the separation period. Then, (S3) and (S4), which are almost identical points on the trajectory (they refer to almost contemporary events), are placed in the part where the feelings of both Scarlett and Rhett are increasing. Finally, the segment (S5) (where they kiss each other for the first time) is roughly placed at the turning point of the trajectory because Scarlett's involvement x_1 at that point is greater than in (S4) but also greater than in (S6) where she slaps him after a kiss.

(S7) is placed at the end of the separation period, while (S11) and (S12) are at the end of the trajectory because they close the love story. The remaining three segments (S8), (S9), and (S10) are placed where they both look happy in such a way that (S9) corresponds to the point of highest involvement of Rhett during the last phase.

In conclusion, we have verified that the sequence (S2)–(S12) can be allocated on the trajectory of Figure 7.5 without clashing with the spontaneous impression one gets when watching the film. Since all this is highly subjective, the reader is invited to go through the same exercise to check if she/he agrees with our proposed allocation of the segments or finds better

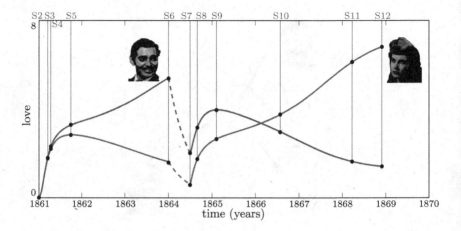

Fig. 7.7 Time evolution of Scarlett's and Rhett's involvements during their love story. Red and blue lines represent the evolution of Scarlett's and Rhett's loves, respectively. The figure points out the chronology of the proposed allocation of the segments S2–S12, which turns out to be consistent with the information available in the film.

solutions.

But we can do even more and perform a second check of the agreement between model and film. For this, we first use the model to plot the time evolution of the feelings, obtaining the two graphs of Figure 7.7, and then determine the chronology of (S2)–(S12) consistent with the allocation proposed in Figure 7.6. This allows the chronology predicted by the model to be checked to see if it is in good enough agreement with that roughly indicated in the film and the novel.

7.4 Conclusions

An original aspect of this chapter is the discussion of the agreement between model and film. As in a context like this quantitative data are not available, the comparison is not carried out in the standard technical way (*e.g.*, by comparing the mean square error between predictions and available measures, or by using more sophisticated but comparable statistical techniques, as done, for example, in Ferrer *et al.* (2012)) but rather by using a much softer approach. Indeed, in the specific application, after the model has produced a trajectory in the love space (see Figure 7.5) representing the predicted time evolution of the involvements of Scarlett and Rhett, eleven chronologically ordered short segments of the film (representing the data!) are analyzed and compared pairwise. This is done with the aim of deriving a clear (though subjective) impression regarding the fact that Scarlett and Rhett's loves increase or decrease from one segment to the next. After this has been done, it is possible to check if the eleven segments can be allocated along the predicted trajectory in a way consistent with the subjective impressions. In the case of "Gone with the Wind" a satisfactory allocation is obtained so that the agreement of the model and the film is somehow verified (see Figure 7.6). However, this is questionable because it is based on our subjective judgment. For this reason, the reader can go through the same exercise to check if he/she agrees with our proposed allocation or not.

It is worth noting that only the first two segments of the film are used to identify the characters of Scarlett and Rhett on which the model is based, while the remaining ones are used to show that the love story described in the film and that predicted by the model are in good agreement. The result is actually astonishing: it shows that the information available after only half an hour is sufficient to predict the evolution and the dramatic end of a story described in a very long film.

Finally, by looking at our work from a slightly different angle, we can say that we have shown in a very technical way that a love story described in a film is highly realistic because it is consistent with the characters of the individuals involved. Although realism is not necessary from a purely artistic point of view, in most cases it is very much appreciated by the audience, to the point of being strategically important in making the film popular. We can then argue that we have perhaps understood, using a mathematical approach, why "Gone with the Wind" has been one of the most successful films of all time.

Chapter 8

Romantic cycles

We have seen in the preceding chapters that two unbiased individuals cannot have recurrent ups and downs in their feelings if they are both secure or both insecure. Since turbulent love stories exist and have attracted a lot of attention (see, for example, Chapters 9, 10, and 13), we investigate in this chapter the existence of periodic romantic regimes using our models.

First we prove, through Bendixon's criterion, that the non-existence of ups and downs just mentioned is a general property of all couples composed of unbiased individuals. This means that we must consider couples with at least one biased individual if we wish to detect periodic romantic regimes. Then, we prove that some insecurity is also needed in the couple for promoting romantic ups and downs. Finally, we discuss the case of two insecure and biased individuals and show, through simulation and bifurcation analysis, that when the bias is smoothly increased, stationary regimes can turn into periodic regimes. Initially, the ups and downs are very small and hence hardly detectable, but then, with a further increase in the bias, the turbulence becomes clearly evident. This type of emergence of periodic regimes is due to a bifurcation known as the Hopf bifurcation in which the cycles are initially very small. Other mechanisms of emergence of periodic regimes, due to other bifurcations, are also possible and are illustrated by reporting the results obtained for a couple where she is secure and biased and he is just the opposite. In this case not only Hopf but also homoclinic bifurcations are possible. Together, they explain why the turbulence of romantic relationships can gradually disappear as the partners get older. Indeed, the Hopf bifurcation predicts that the ups and downs become less and less pronounced, while the homoclinic bifurcation says that the crises or the bursts of love characterizing the romantic relationship remain relevant but become less and less frequent. These abstract examples, as well as the case studies presented in the next two chapters, suggest that consistent bias and insecurity should, in general, promote turbulent romantic relationships.

The notions of Hopf and homoclinic bifurcations, and continuation techniques, are needed for fully appreciating this chapter. Further results can be found in Gragnani *et al.* (1997).

8.1   Introduction

That the behavior of lovers can, in principle, oscillate like the moon (or even be unpredictable like the weather), is a property suggested by the general theory of dynamical systems (see Appendix). Perhaps the first (abstract and elegant) discussion of this idea can be found in Schuster and Sigmund (1981), who use the theory of games and the notion of evolutionarily stable strategy to support the property. On the other hand, there are numerous and well documented love stories characterized by stormy patterns in the feelings, like that identified by Jones (1995) in Petrarch's "Canzoniere", the most celebrated book of love poems of the Western world. The existence of "cycles" with different frequencies in romantic relationships has been recognized in various studies. For example, there are easily detectable 24-hours oscillations in the feelings between two persons if at least one of them has pronounced daily variations in blood pressure. Weekly cycles have also been observed and are obviously due to the difference between weekends and working days. Moon cycles can be also present in young couples because sexual activities are periodically influenced by menstruation. All these cycles are high-frequency cycles which are forced by exogenous factors. On the other hand, the cycles we talk about in this chapter are low-frequency and produced endogenously. From the phenomena encapsulated in the model (oblivion and reactions to love and appeal) one should expect cycles with periods comparable with the time constants of the forgetting processes, *i.e.*, periods ranging from a few months to a few years. For example, the love cycle discovered by Jones in Petrarch's "Canzoniere" is about 4 years long.

8.2   No bias implies no cycles

In the state portraits described in the preceding chapters there are no cycles. This means that the couples studied until now can only converge to stationary romantic regimes. As in Chapters 2–5 we have considered couples composed of unbiased and secure individuals, while in Chapters 6 and 7 the individuals are unbiased and insecure, we can summarize our results by saying that we have shown that unbiased individuals cannot have recurrent ups and downs in their feelings if they are both secure or both insecure. This result is only part of the truth. In fact, two unbiased individuals cannot have cycles. To prove this result consider the model of a couple composed of unbiased individuals

$$\dot{x}_1 = f_1(x_1, x_2) = -\alpha_1 x_1 + R_1^L(x_2) + \gamma_1 A_2$$
$$\dot{x}_2 = f_2(x_1, x_2) = -\alpha_2 x_2 + R_2^L(x_1) + \gamma_2 A_1,$$

and determine its *divergence*, namely

$$\text{div} f = \frac{\partial f_1}{\partial x_1} + \frac{\partial f_2}{\partial x_2} = -\alpha_1 - \alpha_2.$$

Since the divergence does not change sign, Bendixon's criterion (see Appendix) rules out the possibility of existence of limit cycles. In other words, some bias is present in couples with periodic romantic regimes.

8.3 No insecurity implies no cycles

Consider now a generic couple described by the general model

$$\dot{x}_1 = f_1(x_1, x_2) = -\alpha_1 x_1 + R_1^L(x_1, x_2) + R_1^A(x_1, A_2) \tag{8.1}$$
$$\dot{x}_2 = f_2(x_1, x_2) = -\alpha_2 x_2 + R_2^L(x_1, x_2) + R_2^A(x_2, A_1).$$

In this model, individual i is synergic if

$$\frac{\partial R_i^L}{\partial x_i} \geq 0 \qquad \frac{\partial R_i^A}{\partial x_i} \geq 0$$

and at least one of the two inequalities holds with the strict inequality sign, while the opposite inequality signs hold in the case of platonic individuals.

The divergence of the model is

$$\text{div} f = \frac{\partial f_1}{\partial x_1} + \frac{\partial f_2}{\partial x_2} = -\alpha_1 - \alpha_2 + \frac{\partial R_1^L}{\partial x_1} + \frac{\partial R_1^A}{\partial x_1} + \frac{\partial R_2^L}{\partial x_2} + \frac{\partial R_2^A}{\partial x_2},$$

so that its sign is negative if the two individuals are platonic. In that case Bendixon's criterion implies that limit cycles do not exist. This means that permanent ups and downs in the feelings are not possible if the two individuals are platonic. In all other cases, the sign of the divergence is not *a priori* identifiable, so that Bendixon's criterion is of no help. We can, however, prove that if the two individuals are secure, *i.e.*, if

$$\frac{\partial R_1^L}{\partial x_2} > 0 \qquad \frac{\partial R_2^L}{\partial x_1} > 0,$$

cycles cannot be produced through Hopf bifurcations. The proof is by contradiction, *i.e.*, we assume that a Hopf bifurcation exists and show that this is not possible. In fact, recall (see Appendix) that at Hopf bifurcations the Jacobian matrix (evaluated at the equilibrium)

$$J = \begin{vmatrix} J_{11} & J_{12} \\ J_{21} & J_{22} \end{vmatrix} = \begin{vmatrix} \dfrac{\partial f_1}{\partial x_1} & \dfrac{\partial f_1}{\partial x_2} \\[2mm] \dfrac{\partial f_2}{\partial x_1} & \dfrac{\partial f_2}{\partial x_2} \end{vmatrix}$$

must satisfy the following two conditions

$$\mathrm{tr}J = 0 \tag{8.2}$$

$$\det J > 0. \tag{8.3}$$

Thus, if a Hopf bifurcation exists, condition (8.2) is satisfied, *i.e.*,

$$J_{11} + J_{22} = 0,$$

which implies $J_{11}J_{22} \leq 0$. But

$$\det J = J_{11}J_{22} - J_{12}J_{21},$$

where

$$J_{12} = \frac{\partial f_1}{\partial x_2} = \frac{\partial R_1^L}{\partial x_2} \qquad J_{21} = \frac{\partial f_2}{\partial x_1} = \frac{\partial R_2^L}{\partial x_1}.$$

In conclusion, as the two individuals are secure, condition (8.3) cannot be satisfied and this contradicts the assumption.

To prove that limit cycles cannot exist if the two individuals are secure, we should also be able to exclude tangent bifurcations of limit cycles and homoclinic bifurcations, that explain other mechanisms of emergence of limit cycles in second order dynamical systems (see Appendix). A proof of this kind is certainly difficult and is not available. In contrast, exploiting the geometry of the null-clines it is possible, though relatively difficult, to show that limit cycles cannot exist. This proof is not reported here but the interested reader can find it in Gragnani *et al.* (1997).

We can therefore summarize the results discussed in this section by saying that some insecurity is present in couples with periodic romantic regimes.

8.4 A first example of romantic cycle

To show that limit cycles can exist in model (8.1), we present a first simple and abstract example, while a more complex one is discussed in the next section. The couple we consider is composed of two insecure and biased individuals who are synergic in their reaction to appeal. This implies that the model is

$$\dot{x}_1 = -\alpha_1 x_1 + R_1^L(x_2) + (1 + b_1^A B_1^A(x_1))\gamma_1 A_2$$
$$\dot{x}_2 = -\alpha_2 x_2 + R_2^L(x_1) + (1 + b_2^A B_2^A(x_2))\gamma_2 A_1,$$

where the reaction functions R_1^L and R_2^L are first increasing and then decreasing and the bias coefficients b_1^A and b_2^A are positive. Thus, the question is: can this model exhibit limit cycles for suitable values of its parameters?

In agreement with the notation used in the previous chapters, we fix the reaction and synergism functions as follows

$$R_i^L(x_j) = \beta_i k_i x_j \exp(-(k_i x_j)^{n_i})$$
$$B_i^A(x_i) = x_i^{2m_i}/(x_i^{2m_i} + \sigma_i^{2m_i}).$$

Since the reaction to love R_i^L peaks at

$$x_j = \frac{1}{k_i} \sqrt[n_i]{\frac{1}{n_i}},$$

the parameter k_i is a measure of the insecurity of individual i (as in Chapter 6). Now that we have fixed the class of models we wish to study, we show how we can organize our numerical analysis in order to detect if limit cycles can exist for suitable values of the bias coefficients.

First we fix the parameters, for example at the following reference values

$$\alpha_1 = 0.36 \quad k_1 = 0.08 \quad n_1 = 1 \quad \beta_1 = 0.75$$
$$A_1 = 0.1 \quad \sigma_1 = 1 \quad m_1 = 4 \quad \gamma_1 = 1$$

$$\alpha_2 = 0.2 \quad k_2 = 1.5 \quad n_2 = 4 \quad \beta_2 = 10.66$$
$$A_2 = 0.1 \quad \sigma_2 = 1 \quad m_2 = 4 \quad \gamma_2 = 1 \,.$$

while we leave the bias coefficients b_1^A and b_2^A free, at least for the moment. For these parameter values the reaction and the bias functions are as in Figure 8.1.

Then we look for bias coefficients for which the couple has a stable equilibrium. This is easy because we already know that unbiased couples cannot have limit cycles. Thus, we can fix $b_1^A = 0$ and $b_2^A = 0$ and determine by simulation the corresponding equilibrium.

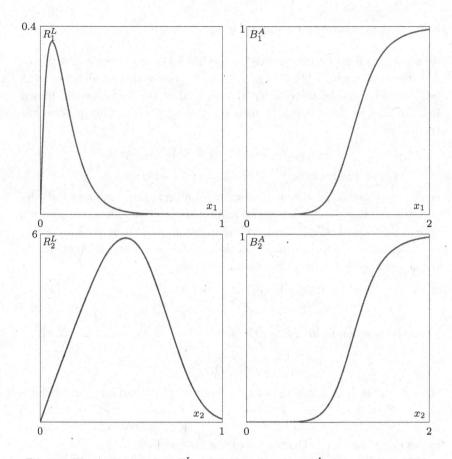

Fig. 8.1   The reactions to love R_i^L and the bias functions B_i^A of the two individuals.

Then we can increase, step by step, one of the two bias coefficients and repeat the simulations at each step hoping to detect a cycle. This point is critical because, in agreement with the above discussion, it is not guaranteed that by increasing the bias coefficient a cycle will emerge. In the present case, no cycle is found if the bias b_2^A is increased, while a cycle emerges for

$$b_1^A = b_1^{A^*} = 2.93 \qquad (8.4)$$

if the first bias coefficient is gradually increased. The cycle is very small for values just above the critical value (8.4), but then it quickly becomes larger and its shape and period continue to vary with the parameter b_1^A. Figure 8.2 is the most effective representation of a Hopf bifurcation (see

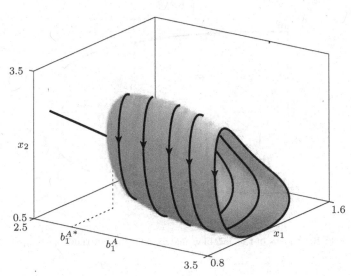

Fig. 8.2   The attractor of the model for different values of the bias coefficient b_1^A.

Appendix). It shows that a stable equilibrium becomes unstable at the threshold value b_1^{A*} of the parameter and is surrounded, there after, by a stable limit cycle that is initially very small.

Since for $b_1^A = b_1^{A*}$ the equilibrium loses stability, the two eigenvalues of the Jacobian matrix evaluated at the equilibrium are purely imaginary ($\lambda_{1,2} = \pm i\omega$). This means that the Hopf bifurcation can also be represented by plotting in the complex plane the so-called *root locus* of the system, as performed in Figure 8.3. This locus shows that for $b_1^A < b_1^{A*}$ the two eigenvalues are complex conjugate and have a negative real part (*i.e.*, the equilibrium is a stable focus), while for $b_1^A > b_1^{A*}$ they are still complex conjugate but have a positive real part (unstable focus). In other words, the eigenvalues λ_1 and λ_2 cross the imaginary axis at the points $\pm i\omega$ where ω is the frequency of the cycle emerging at the bifurcation ($\omega = 2\pi/T$, where T is the period of the emerging cycle).

Now that we have found a Hopf bifurcation by varying the bias coefficient b_1^A of the first individual and assuming that the second one is unbiased ($b_2^A = 0$), we can extend our analysis by repeating the same computations for a series of positive values of b_2^A. The result is a curve in the two-dimensional space of the parameters (b_1^A, b_2^A) separating the region where the couple tends to a stationary regime from the region where it tends to a periodic regime. This curve, indicated by H in Figure 8.4, is called the Hopf bifurcation curve.

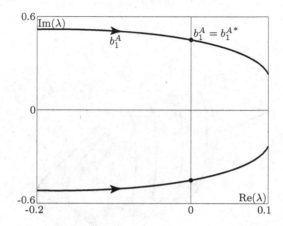

Fig. 8.3   Root locus of the Jacobian matrix pointing out the Hopf bifurcation.

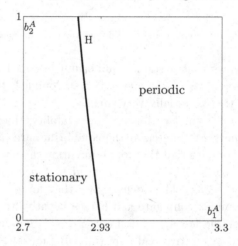

Fig. 8.4   Hopf bifurcation curve separating stationary from periodic romantic regimes.

Before closing this section, it is important to say that if the Hopf bifurcation curve were really obtained through the successive simulations we have just described, the analysis would be terribly long and inaccurate. This is because in the vicinity of a bifurcation the convergence toward the stable equilibrium or toward the stable cycle is so slow that it becomes practically impossible to distinguish if we are in the stationary or periodic region of Figure 8.4. Fortunately, this difficulty can be circumvented by using a numerical continuation method for automatically producing the entire Hopf bifurcation curve (see Appendix). The method proceeds as follows. Once a first point of the Hopf bifurcation curve has been obtained (see point $b_1^A = 2.93$, $b_2^A = 0$), the entire curve is constructed by producing iteratively a dense sequence of points on the curve. Each new point is determined from the last one by suitably varying the two parameters of a small amount in such a way that the property identifying the bifurcation (here the anni- hilation of the real part of the eigenvalues) is conserved along the curve. As mentioned in the Appendix, the software available today for performing the continuation of bifurcation curves is very effective. For example, the curve H in Figure 8.4 is automatically produced in about one second on a standard PC.

Continuation techniques can also be used to detect the influence of other parameters. For example, Figure 8.5 shows that the region where the ro- mantic regime is periodic becomes smaller if the insecurity coefficient k_1 of the first individual is decreased. Each bifurcation curve in Figure 8.5 is produced starting from a point $(\bar{b}_1^A, \bar{b}_2^A)$ on a previous curve, say the one corresponding to $k_1 = \bar{k}_1$ and by performing the continuation of the Hopf bifurcation with respect to b_1^A and k_1, with b_2^A frozen at the value $\bar{b}_2^A$. Then, once a point $(\bar{\bar{b}}_1^A, \bar{b}_2^A)$ on the curve corresponding to a value $\bar{\bar{k}}_1$ of k_1 is obtained, k_1 is frozen at $\bar{\bar{k}}_1$ and the Hopf bifurcation is continued with respect to b_1^A and b_2^A starting from point $(\bar{\bar{b}}_1^A, \bar{b}_2^A)$. Thus, a family of Hopf bifurcation curves corresponding to different values of k_1 is produced recursively with quite a limited computational effort.

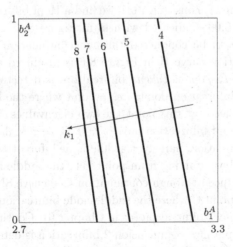

Fig. 8.5 Influence of the insecurity k_1 of the first individual on the Hopf bifurcation curve. The value of $100k_1$ is reported on each curve.

8.5 A second example of romantic cycle

The example discussed in the previous section is relatively simple because only one bifurcation is present. In other couples, like the one described in this section, the analysis is more complex because there are two or more bifurcations. The couple we consider here is the first one for which romantic cycles have been discovered (Gragnani *et al.*, 1997). In this couple she is secure and biased, while he is just the opposite. Thus, assuming that she is synergic only in her reaction to appeal, the model is

$$\begin{aligned}
\dot{x}_1 &= -\alpha_1 x_1 + R_1^L(x_2) + (1 + b_1^A B_1^A(x_1))\gamma_1 A_2 \\
\dot{x}_2 &= -\alpha_2 x_2 + R_2^L(x_1) + \gamma_2 A_1,
\end{aligned} \tag{8.5}$$

where b_1^A is positive and $R_1^L(x_2)$ is increasing while $R_2^L(x_1)$ is first increasing and then decreasing.

The model can be analyzed following the same approach as that used in the previous section (for details, see Gragnani *et al.* (1997)). First we fix the functional forms of the reactions to love and the reference parameter values, leaving two of them free, for example, her bias coefficient b_1^A and his appeal A_2. Then we select the two free parameters in such a way that the couple has a stable stationary regime. As in the previous example, this can easily be done because we know that for low bias coefficients the couple cannot have turbulent behaviors (see, for example, point $b_1^A = 1.4$, $A_2 = 0.098$ indicated with ★ in Figure 8.6). Then we increase the bias coefficient, keeping A_2 constant, until we find a Hopf bifurcation (see point $b_1^A = 1.5$, $A_2 = 0.098$ indicated with ■ in Figure 8.6). From this point on, the analysis can be conducted through continuation. Thus, the entire Hopf bifurcation curve H in Figure 8.6 is produced and a particular point BT, called *Bogdanov-Takens* bifurcation, is detected on it. This is a codimension-2 bifurcation point, *i.e.*, a point where the Hopf bifurcation is degenerate. Indeed, at this point the two eigenvalues $\lambda_{1,2} = \pm i\omega$ associated with the Hopf bifurcation have $\omega = 0$. This reveals (see Appendix) that a second bifurcation, namely, a saddle-node bifurcation, is also present in the system. Thus, starting from point BT, the saddle-node bifurcation curve can be produced through continuation (see curve SN in Figure 8.6), ending at a cusp-point C where the saddle-node bifurcation actually degenerates into a pitchfork bifurcation (as in Chapter 6). Continuing the second branch of the cusp a new codimension-2 bifurcation is detected at point D in Figure 8.6. At this point two homoclinic bifurcation curves (p and q in Figure 8.6) merge, and curve p terminates at the Bogdanov-Takens point

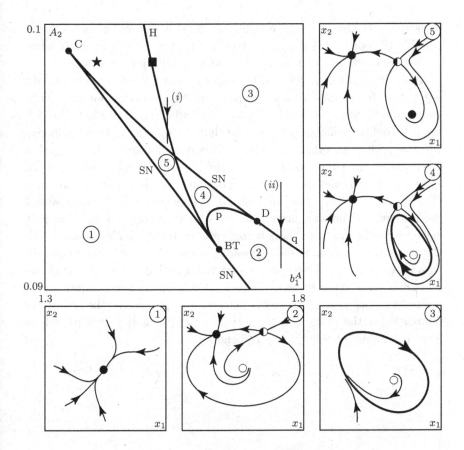

Fig. 8.6 The complete bifurcation diagram of model (8.5) for the functions and parameters values used in Gragnani *et al.* (1997) is reported in the main panel. State portraits in the five regions ①, ②, ..., ⑤ are qualitatively sketched in the small panels.

BT. Figure 8.6 is the complete bifurcation diagram of model (8.5) with respect to the two parameters b_1^A and A_2. This diagram points out the full catalog of behaviors of model (8.5). The considered space of parameters is partitioned into five subregions ①, ②,..., ⑤ in which the state portraits are as those sketched in Figure 8.6. In subregions ④ and ⑤, *i.e.*, in the cusp region, the couple has two alternative attractors, as in the cusp region discussed in Chapter 6. However, the novelty is that in one of the two subregions (namely ④) one of the two attractors is a limit cycle. This means that a couple with parameter values in region ④ can converge either to a stationary romantic regime or to a periodic romantic regime depending upon the past history of the couple. In all other regions (namely ①, ②, and ③) the couple has a unique attractor, namely a cycle for high values of the bias coefficient (subregion ③) and an equilibrium otherwise.

Interestingly, Figure 8.6 can be used to point out, in a basic way, the role played by age in turbulent couples. For this, consider, for example, a young couple with $b_1^A = 1.54$ and $A_2 = 0.097$, which, as shown in Figure 8.6, has a turbulent behavior, and assume that the appeal A_2 is slowly fading with age, while all other parameters remain unchanged. Thus, the slow evolution of the couple occurs along the vertical straight line (i) in the space of the parameters. When approaching the Hopf bifurcation curve H along path (i) the cycle shrinks and disappears on H. In other words, the romantic turbulence present in the couple smoothly disappears with age because the ups and downs become gradually less and less pronounced. But the turbulence can also disappear through a different pattern, namely, with crises or bursts of interest (characterizing turbulent romantic regimes) becoming less and less frequent. This is what happens if the evolution with age occurs along path (ii) in Figure 8.6 because approaching the homoclinic bifurcation q the cycle is characterized by longer and longer periods of almost stationary behavior (see Appendix).

Chapter 9

Kathe and Jules

In this chapter we study the love story involving Helen Grund—a brilliant and charming journalist from Berlin—and Franz Hessel—a profound but shy German writer. Their relationship is characterized by ups and downs: they married and divorced twice and lived again together after the second divorce.

Following the standard approach of psychoanalysis, we first identify the main psychological traits of the two lovers from the best available source of information, the novel "Jules et Jim" written by Henry Pierre Roché, a friend of Franz Hessel. The novel, popularized by the film of the same name directed by François Truffaut, is quite reliable because it is autobiographic.

In the novel, as well as in the film, Helen and Franz are Kathe and Jules, respectively (see Figure 9.1). Jules is secure and platonic (his reaction to

Jeanne Moreau   Oskar Werner

Kathe   Jules

Fig. 9.1   Shots of the film "Jules et Jim".

Kathe's appeal is diminished the more in love with her he is). Kathe is a passionate woman: she is insecure (because she reacts negatively to the platonic nature of Jules) and synergic in her reaction to appeal. Insecurity and synergism—the necessary ingredients for the emergence of amorous cycles—are therefore present in the couple. Fixing the parameters of the model at realistic reference values, the love story evolves toward a romantic cycle with a period of approximately 4 years. This and other details are consistent with the information available from the novel. However, as the reference parameter values are largely subjective, the results of a simple bifurcation analysis are also shown to discuss the reliability of the discovered cycle. The conclusion is that the existence of the cycle depends critically on some parameters, in particular Jules' platonicity. This somehow supports the common belief, already discussed in Chapter 5, that small details can have great consequences in love affairs.

No new mathematical prerequisites are needed for reading this chapter. Extra details can be found in Dercole and Rinaldi (2014).

9.1   The story

"Jules et Jim" is the first novel by Henri-Pierre Roché, published in 1953 (Roché, 1953) when he was already 74. It is an autobiographical novel describing the love story involving Helen Grund (Kathe), her husband Franz Hessel (Jules), and his best friend Roché (Jim). The story begins a few years before the World War I in Paris, where it ends about 20 years later.

Roché's novel is interesting for two reasons. First because, being autobiographical, it is a reliable source of information, thus allowing us to validate a mathematical model against a real love story. Second because it conveys the central idea of Roché's philosophy, namely, that we should not try to possess or constrain the people we love but leave them free to engage in other relationships. This anti-bourgeois ideology, later revisited in the "free love" and "sexual liberation" movements popular in the sixties and seventies, is important for our purposes because it suggests we should model the couple Kathe-Jules in the virtual absence of Jim. Similarly, in the next chapter, we study the couple Kathe-Jim in the virtual absence of Jules. In other words, Kathe, described by Roché as a follower of the free love ideology, is living two parallel love stories with no particular internal conflict, while the deep friendship between Jules and Jim attenuates possible jealousies. As explicitly admitted by Roché in the following quotes (from the English translation of Roché's novel (Roché, 1953)), the two stories basically develop independently:

> *In her mind, each lover was a separate world, and what happened in one world was no concern of the others* (p.108)

> *In twenty years Jim and he had never quarrelled. Such disagreements as they did have they noted indulgently* (p.237)

The importance of the free-love hypothesis is discussed in Chapter 15, where the weak but inevitable interactions in the triangular relationship are taken into account.

The story of "Jules et Jim" became famous worldwide after the success of the 1961 film of the same name—a celebrated masterpiece of the French *Nouvelle Vague* directed by François Truffaut. For our modeling purposes, the novel and the film are actually interchangeable, as the film screenplay follows the novel almost word by word. The few, but extremely interesting, innovations added by Truffaut are discussed at the end of Chapter 15.

9.2 The Kathe-Jules model

We now build the model of the couple Kathe-Jules, by providing excerpts from Roché's novel to support our choices. In agreement with our standard notation, we indicate Kathe's feeling for Jules with x_1 and Jules' feeling for her with x_2.

The main peculiarity of Jules is that he is platonic:

> *Really, Jules is happy, in his own way, and just wants things to go on. He's seeing you often, in idyllic circumstances, and he's living on hope* (p.24)

He therefore reduces his reaction to Kathe's appeal when he is more in love with her, that is, his reaction is attenuated by the factor $(1 - P(x_2))$, where $P(x_2)$ is Jules' platonicity (see Figure 9.2). In accordance with his platonic nature, Jules is a secure lover, and assuming his reaction to love $R_2^L(x_1)$ to be linear (*i.e.*, $R_2^L(x_1) = \beta_2 x_1$, see Figure 9.2), the equation regulating his feeling is

$$\dot{x}_2 = -\alpha_2 x_2 + \beta_2 x_1 + (1 - P(x_2))\gamma_2 A_1. \tag{9.1}$$

Kathe is a passionate woman and, though charmed by Jules, she is annoyed by his platonic nature:

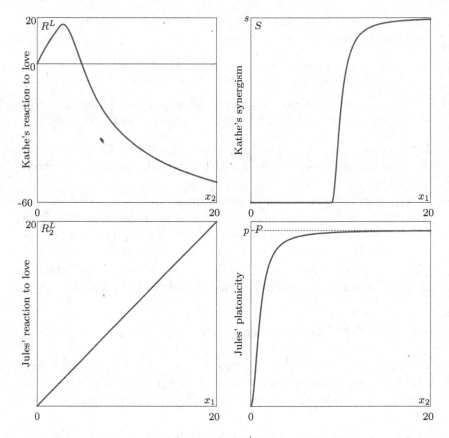

Fig. 9.2 The reactions to love R^L and R_2^L of Kathe and Jules, Kathe's synergism S, and Jules' platonicity P. See Table 9.1 for the analytical expressions of these functions and for the values of their parameters.

> *But then, he wasn't the husband she needed, and she wasn't the woman to bear that (p.89)*

> *She had been drawn by his mind, his gift of fantasy. But she needed, in addition to Jules, a male of her own sort (p.90)*

For this reason, her reaction $R^L(x_2)$ to Jules' love is of the insecure type. Moreover, with Kathe being an enthusiastic person, her reaction to Jules' appeal is amplified by the factor $(1 + S(x_1))$, where $S(x_1)$ is her synergism (see top panels in Figure 9.2). In conclusion, Kathe's equation is

$$\dot{x}_1 = -\alpha_1 x_1 + R^L(x_2) + (1 + S(x_1))\,\gamma_1\,A_2. \tag{9.2}$$

Nonlinear functions (specified for non-negative feelings)

Kathe	$R^L(x_2) = \beta_1 \dfrac{x_2}{1+x_2/\sigma^L} \cdot \begin{cases} \dfrac{1-((x_2-\tau^I)/\sigma^I)^2}{1+((x_2-\tau^I)/\sigma^I)^2} & \text{if } x_2 \geq \tau^I \\ \qquad\qquad 1 & \text{if } x_2 < \tau^I \end{cases}$	reaction to love
	$S(x_1) = \begin{cases} s\,\dfrac{((x_1-\tau^S)/\sigma^S)^2}{1+((x_1-\tau^S)/\sigma^S)^2} & \text{if } x_1 \geq \tau^S \\ \qquad\quad 0 & \text{if } x_1 < \tau^S \end{cases}$	synergism
Jules	$P(x_2) = \begin{cases} p\,\dfrac{((x_2-\tau^P)/\sigma^P)^2}{1+((x_2-\tau^P)/\sigma^P)^2} & \text{if } x_2 \geq \tau^P \\ \qquad\quad 0 & \text{if } x_2 < \tau^P \end{cases}$	platonicity

Parameters

Kathe	forgetting	α_1	$= 2\ [\text{yrs}^{-1}]$	forgetting coefficient
	reaction to love	β_1	$= 8\ [\text{yrs}^{-1}]$	reaction coefficient to love
		σ^L	$= 10$	sensitivity of reaction to love
		τ^I	$= 2.5$	insecureness threshold
		σ^I	$= 10.5$	sensitivity of insecureness
	reaction to appeal	γ_1	$= 1\ [\text{yrs}^{-1}]$	reaction coefficient to appeal
	synergism	s	$= 2$	maximum synergism
		τ^S	$= 9$	synergism threshold
		σ^S	$= 1$	sensitivity of synergism
	appeal	A_1	$= 20$	appeal
Jules	forgetting	α_2	$= 1\ [\text{yrs}^{-1}]$	forgetting coefficient
	reaction to love	β_2	$= 1\ [\text{yrs}^{-1}]$	reaction coefficient to love
	reaction to appeal	γ_2	$= 0.5\ [\text{yrs}^{-1}]$	reaction coefficient to appeal
	platonicity	p	$= 1$	maximum platonicity
		τ^P	$= 0$	platonicity threshold
		σ^P	$= 1$	sensitivity of platonicity
	appeal	A_2	$= 4$	appeal

Table 9.1   Nonlinear functions and reference parameter values of the Kathe-Jules model.

The model of the couple Kathe-Jules is therefore composed of equations (9.1, 9.2), with functions R^L, S, and P as in Figure 9.2. Of course, we must assign reasonable values to all parameters, taking into account all possible indications emerging from the novel. For example, we take Kathe's appeal A_1 to be greater than Jules appeal A_2 because she is definitely more fascinating. Similarly, we assume she forgets faster than him, $\alpha_1 > \alpha_2$, being the more unstable in the couple. Of course, the specific values we have assigned to the parameters (called reference parameter values, see Table 9.1) remain rather arbitrary and based on our subjective interpretations.

9.3   Analysis and results

Model (9.1, 9.2) can be simulated to compute the time evolution of the feelings of Kathe and Jules. The outcome of the simulation starting from the usual initial condition of indifference, $x_1(0) = x_2(0) = 0$, is reported in Figure 9.3. It shows that the love story tends toward a periodic regime with a period of about 4 years, more precisely, 3 years and 10 months.

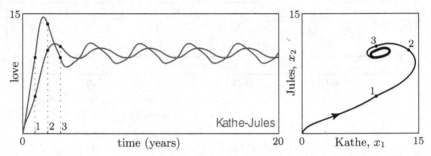

Fig. 9.3   The love story predicted by the Kathe-Jules model.

The result can be constructed day by day (with no noticeable difference at the scale of the figure) by using the trivial discretization $\dot{x}_i = x_i(t+1) - x_i(t)$, $i = 1, 2$. When this is done, the continuous-time model (9.1, 9.2) is substituted by the discrete-time model

$$x_2(t+1) = x_2(t) - \alpha_2\, x_2(t) + \beta_2\, x_1(t) + (1 - P(x_2(t)))\, \gamma_2\, A_1 \quad (9.3)$$

$$x_1(t+1) = x_1(t) - \alpha_1\, x_1(t) + R^L(x_2(t)) + (1 + S(x_1(t)))\, \gamma_1\, A_2 \quad (9.4)$$

where the time unit is "one day" and the value of parameters α_1, β_1, γ_1, α_2, β_2, γ_2 (with unit [yrs^{-1}] in Table 9.1) must be divided by 365 in view of the time scaling. For example, starting with $x_1(0) = x_2(0) = 0$ at day 0, and using equations (9.3, 9.4) with $t = 0$ to compute the values of the two feelings at day 1, we obtain $x_1(1) = \gamma_1\, A_2$ and $x_2(1) = \gamma_2\, A_1$. This clearly shows that only appeal matters at the beginning of a love story, as feelings are still latent. To go on to the next day, time is increased by one unit and equations (9.3, 9.4) written for $t = 1$ are used to compute the feelings at day $t = 2$. Note that forgetting and reaction to love are also involved now. Repeating the same operations for $t = 2, 3, \ldots$, we can compute the feelings of Kathe and Jules at day $3, 4, \ldots$, and continue in this way for years to produce the time series of Figure 9.3. In the figure, the points indicated by 1, 2, and 3 represent the feelings of Kathe and Jules at the end of the first, second, and third years of their relationship.

According to the model, Kathe and Jules are always positively involved. This means that there is no need to specify the functions in Figure 9.2 for negative values of the feelings. The love story does not reach a plateau, but is instead characterized by recurrent ups and downs that soon become periodic (see Figure 9.3). At the beginning of their relationship, Kathe and Jules are increasingly involved, until Kathe has the first inversion in her trend (the first local maximum of x_1). This and the following inversions of Kathe's involvement are naturally perceived with stress and fear by Jules, as explicitly noted by Roché:

> The danger was that Kate would leave. She had done it once already, half a year ago, and it had looked as if she didn't mean to return... She was full of stress again, Jules could feel that she was working up for something (p.89)

This passage is also emphasized by Truffaut, who shows in a scene of the film Jules confessing to Jim that he feels Kathe is ready to leave him for the second time.

The chapter ends with a parametric analysis of model (9.1, 9.2) aimed at:

(i) checking whether the limit cycle shown in Figure 9.3 is the result of a particular choice of some critical parameters, or its existence is guaranteed for reasonable parameter perturbations from the reference values of Table 9.1;

(ii) studying the sensitivity of the cycle characteristics, such as amplitude and period, with respect to the most critical parameters;

(iii) analyzing the model behavior for parameter values slightly different from those guaranteeing the existence of the limit cycle.

To answer question (i) we can perform a bifurcation analysis of model (9.1, 9.2) by varying one parameter at a time, keeping all others at their reference values. However, more interesting and informative is the analysis with respect to pairs of parameters, producing two-dimensional bifurcation diagrams. One of these bifurcation diagrams is shown in Figure 9.4, where the two parameters are Kathe's maximum synergism s and Jules' maximum platonicity p. In the diagram, two bifurcation curves, indicated by H and TC, partition the parameter space in three regions. For parameters in regions ① and ②, the model has only one attractor—an equilibrium or a cycle, respectively—while in region ③ there are two alternative attractors—an equilibrium and a cycle, as indicated by the state portraits of the feelings

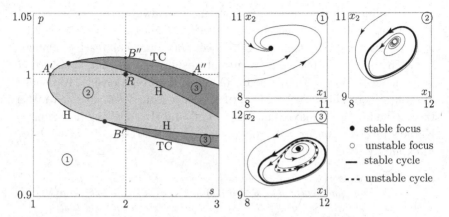

Fig. 9.4   Two-dimensional bifurcation diagram with respect to Kathe's maximum synergism s and Jules' maximum platonicity p. H and TC are Hopf and tangent of limit cycles bifurcations, respectively. For pairs (s, p) in the three regions, ①, ②, and ③ the state portraits are as in the side panels.

(x_1, x_2) reported in the side panels of the figure. Point R corresponds to the reference values of s and p and belongs to region ②. The state portrait ② is indeed the model behavior at R and shows the stable limit cycle that has already been discussed in Figure 9.3.

If we start from point R and gradually decrease Kathe's maximum synergism keeping Jules' maximum platonicity constant, that is, if we move from R toward A', the cycle gradually shrinks and finally becomes a stable equilibrium at point A'. Thus, the transition from region ② to region ① at point A' is a supercritical (non-catastrophic) Hopf bifurcation (see Appendix). In words, the periodic romantic regime is smoothly substituted by a stationary regime if Kathe's synergism becomes too weak. This is consistent with the previous chapter, where we saw that a certain degree of synergism is necessary to obtain romantic ups and downs. In contrast, if we start from point R and increase Kathe's maximum synergism, the cycle disappears only when the bifurcation curve TC is crossed at point A'', where the stable and unstable cycles coexisting in region ③ collide (tangent bifurcation of limit cycles, see Appendix). In other words, this time the periodic romantic regime disappears suddenly and is substituted by a stationary romantic regime. The same kind of discontinuous (catastrophic) transitions are obtained if, starting from point R, we vary only Jules' maximum platonicity because the two extreme points B' and B'' of the region of existence of the cycle are both on the tangent bifurcation TC.

The property pointed out in Figure 9.4, namely, that the cycle disappears perturbing Kathe's synergism and Jules' platonicity in both directions, basically holds true for all parameters of the model. This simply means that the psycho-physical traits of Kathe and Jules should not be too weak or too strong to produce romantic ups and downs. Figure 9.4 shows, however, that there are parameters, like Kathe's maximum synergism, for which the cycle persists for quite large perturbations, and others, like Jules' maximum platonicity, for which the cycle disappears for relatively small perturbations (segment RB'' is only 1.5% of the reference value). Parameters for which the cycle is lost by 5% perturbations in both directions are called "critical" in the following. Although, in general, the presence of critical parameters might bring into question the reliability of a model, in the case of Kathe and Jules the ups and downs in their relationship are so well described by Roché that we can conclude that the peculiarity of this love story lies in the couple itself and is therefore reflected by the presence of some critical parameters in the model.

To detect all critical parameters, the bifurcation analysis was systematically performed with respect to all parameters. Two extra bifurcation diagrams are shown in Figure 9.5. In the first, involving only Kathe's characteristics, one parameter is critical—Kathe's synergism threshold τ^S—while the other is not—sensitivity σ^L of her reaction to love. In the second diagram, involving only Jules' characteristics, again one parameter is

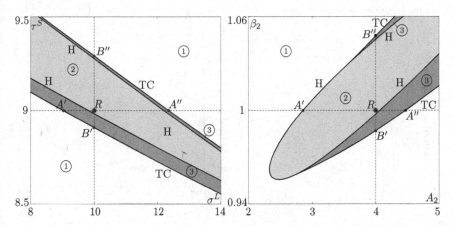

Fig. 9.5 Two other two-dimensional bifurcation diagrams. The parameters are: Kathe's sensitivity of reaction to love σ^L and Kathe's synergism threshold τ^S, left panel; Jules' appeal A_2 and Jules' reaction coefficient to love β_2, right panel. The parameters on the two vertical axis (namely, τ^S and β_2) are critical.

critical—Jules' reaction coefficient to love β_2—while the other is not—Jules' appeal A_2. The result of this systematic analysis is that Kathe has only 1 out of 10 critical parameter (τ^S), while Jules has 3 out of 7 critical parameters (forgetting coefficient α_2, reaction coefficient to love β_2, maximum platonicity p). We can therefore conclude that Jules is the more unusual character of the couple in the sense that many other platonic lovers would not have experienced the ups and downs that he did with Kathe.

To answer question (ii), we have focused on two important characteristics of the cycle: its amplitude (measured by the ratios $x_i^{\max}/x_i^{\min}$ along the cycle) and its period. This can be done by continuation, the same numerical technique used for bifurcation analysis (see Appendix). The results obtained for the most critical parameter—Jules' platonicity p—are shown in Figure 9.6, where the curves are normalized to their reference values. The figure shows that the parameter has a weak effect on the cycle characteristics in a large portion of the narrow range guaranteeing the existence of the cycle. As the same result also holds true for all other critical parameters, we can conclude that almost any value of each critical parameter in the range of existence of the cycle satisfactorily represents the Kathe-Jules love story.

Finally, to answer question (iii), we have checked the model behavior corresponding to a 10% perturbation of each of the four critical parameters.

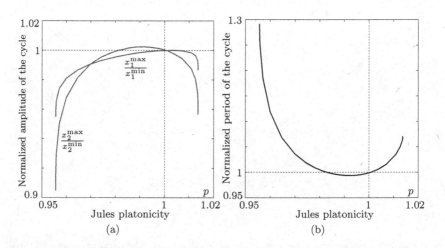

Fig. 9.6   Dependence of the amplitude (a) and of the period (b) of the cycle upon the most critical parameter, Jules' platonicity p. The curves are normalized with respect to amplitude and period of the reference cycle.

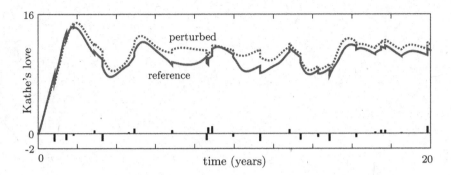

Fig. 9.7   Time evolutions of Kathe's love for reference and 10% perturbed value of the most critical parameter (Jules' maximum platonicity, p). The two time series are obtained by adding to the feelings produced by the model the sequence of random shocks shown at the bottom.

The result is shown in Figure 9.7 for the perturbation of Jules' maximum platonicity p. Interestingly, in the realistic context in which the couple is subject to small shock perturbations (random in both frequency and intensity), remarkable ups and downs remain present in the time-span of the love story, even if the parameters are set outside the region guaranteeing the existence of the periodic regime.

Chapter 10

Kathe and Jim

In this chapter we study the love story between Helen Grund and Henry-Pierre Roché, the friend of her husband. As in the previous chapter, we use the novel "Jules et Jim" as a reliable source of information to understand the characters of the two lovers (Kathe and Jim, see Figure 10.1). Jim is insecure, as all "Don Juans" are to avoid deep involvements and unbiased. Kathe is secure and synergic. Thus, insecurity and synergism—the ingredients that have been shown to be necessary for the emergence of romantic turbulence—are present in the couple. With the parameters of the model fixed at realistic reference values, simulations show that the couple is characterized by a periodic romantic regime. The period of the cycle, as well as other details, are in agreement with the novel. Moreover, the results of a simple bifurcation analysis support the credibility of the identified romantic cycle.

Jeanne Moreau　　　　　Henri Serre

Kathe　　　　　Jim

Fig. 10.1　Shots of the film "Jules et Jim".

No new mathematical prerequisites are needed for reading this chapter. Extra details can be found in Dercole and Rinaldi (2014).

10.1   The Kathe-Jim model

As done in the previous chapter for the couple Kathe-Jules, we now use Roché's ideology of "free love" to build the model of the virtual couple Kathe-Jim, as if Jules were absent. We base our modeling choices on the careful analysis of Roché's novel (Roché, 1953) and report excerpts (from the English translation) as illustrative supporting examples. The variables x_1 and x_2 are Kathe's and Jim's feelings, respectively.

The main characteristic of Jim (see du Toit (2015) for a detailed analysis of the relationship between Helen Grund and Henri-Pierre Roché) is to be insecure, as all "Don Juans" are to avoid deep involvements:

> *'Oh, when,' she said to him one day,—'when are you going to*
> *stop giving me bits of yourself and give me everything?'* (p.207)

Thus, his reaction R^L to Kathe's love is nonlinear and shaped as in Figure 10.2, right panel. Jim being an unbiased character, his equation is

$$\dot{x}_2 = -\alpha_2\,x_2 + R^L(x_1) + \gamma_2\,A_1, \qquad (10.1)$$

where α_2 is Jim's forgetting coefficient and A_1 is Kathe's appeal.

Kathe is secure in her relationship with Jim and synergic in her reaction to appeal, as already described in the previous chapter. Kathe's equation is therefore

$$\dot{x}_1 = -\alpha_1\,x_1 + \beta_1 x_2 + (1 + S(x_1))\,\gamma_1\,A_2, \qquad (10.2)$$

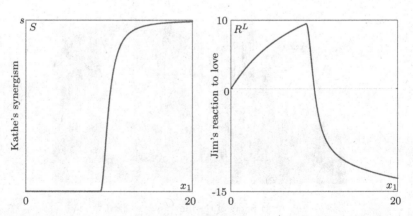

Fig. 10.2   Kathe's synergism function, left panel. Jim's reaction to Kathe's love–typical of an insecure individual, right panel. See Table 10.1 for the analytical expressions and reference parameters values.

Nonlinear functions (specified for non-negative feelings)

Kathe	$S(x_1) = \begin{cases} s\dfrac{((x_1 - \tau^S)/\sigma^S)^2}{1 + ((x_1 - \tau^S)/\sigma^S)^2} & \text{if } x_1 \geq \tau^S \\ \qquad 0 & \text{if } x_1 < \tau^S \end{cases}$	synergism
Jim	$R^L(x_1) = \beta_2 \dfrac{x_1}{1 + x_1/\sigma^L} \cdot \begin{cases} \dfrac{1 - ((x_1 - \tau^I)/\sigma^I)^2}{1 + ((x_1 - \tau^I)/\sigma^I)^2} & \text{if } x_1 \geq \tau^I \\ \qquad 1 & \text{if } x_1 < \tau^I \end{cases}$	reaction to love

Parameters

Kathe	forgetting	α_1	=	2 [yrs^{-1}]	forgetting coefficient
	reaction to love	β_1	=	1 [yrs^{-1}]	reaction coefficient to love
	reaction to appeal	γ_1	=	1 [yrs^{-1}]	reaction coefficient to appeal
	synergism	s	=	2	maximum synergism
		τ^S	=	9	synergism threshold
		σ^S	=	1	sensitivity of synergism
	appeal	A_1	=	20	appeal
Jim	forgetting	α_2	=	2 [yrs^{-1}]	forgetting coefficient
	reaction to love	β_2	=	2 [yrs^{-1}]	reaction coefficient to love
		σ^L	=	10	sensitivity of reaction to love
		τ^I	=	9	insecureness threshold
		σ^I	=	1	sensitivity of insecureness
	reaction to appeal	γ_2	=	1 [yrs^{-1}]	reaction coefficient to appeal
	appeal	A_2	=	5	appeal

Table 10.1   Nonlinear functions and reference parameter values of the Kathe-Jim model.

where α_1 is Kathe's forgetting coefficient, A_2 is Jim's appeal, and S is Kathe's synergism (see Figure 10.2, left panel).

In conclusion, the Kathe-Jim model is composed of equations (10.1, 10.2), with the functions R^L and S as in Table 10.1 and Figure 10.2. The parameters are fixed at reasonable values. For example, Jim being a charming "viveur," his appeal A_2 is assumed to be greater than Jules' appeal, but smaller than Kathe's one (compare with Table 9.1). Since Kathe follows the free love ideology and Jim is a real "Don Juan", they forget rather quickly their past involvements, so that their forgetting coefficients can reasonably be assumed to be greater than that of Jules (compare with Table 9.1). In any case, the parameters are rather arbitrary because they are based on our personal interpretations.

10.2   Analysis and results

Once all parameters are fixed at the values suggested in Table 10.1, the Kathe–Jim model (10.1, 10.2) can be simulated from the initial condition of indifference, $x_1(0) = x_2(0) = 0$, to obtain the time evolution of the feelings of Kathe and Jim. The result is reported in Figure 10.3, where points 1, 2, 3 indicate the feelings of Kathe and Jim at the end of the first, second, and third years of their relationship. As done in the previous chapter, the simulation is carried out day by day by using the discrete-time model

$$x_2(t+1) = x_2(t) - \alpha_2\,x_2(t) + R^L(x_1(t)) + \gamma_2\,A_1 \qquad (10.3)$$

$$x_1(t+1) = x_1(t) - \alpha_1\,x_1(t) + \beta_1\,x_2(t) + (1 + S(x_1(t)))\,\gamma_1\,A_2 \quad (10.4)$$

where the values of parameters α_i, β_i, γ_i (with unit [yrs^{-1}]) are obtained by dividing by 365 the values reported in Table 10.1.

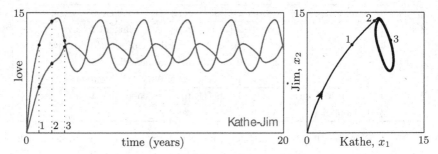

Fig. 10.3   The love story predicted by the Kathe-Jim model.

The two lovers are always positively involved (so that there is no need to specify the functions R^L and S for negative values of the feelings). After an initial positive trend, the love story tends in a few years toward a periodic regime with period of about 3 years and 4 months. The first to invert the positive trend is Jim who, being insecure, refuses too deep involvements:

> *He himself was incapable of living for months at a time in close contact with Kate, it always brought him into a state of exhaustion and involuntary recoil which was the cause of their disasters* (p.189)

As a consequence, the cycle develops clockwise in the plane (x_1, x_2) of the two feelings, as shown in Figure 10.3. Comparing with Figure 9.3, we can note that Jim's oscillations are wider than those of Jules, as explicitly noted by Roché:

Jim was easy for her to take, but hard to keep. Jim's love drops
to zero when Kate's does, and shoots up to a hundred with hers.
I never reached their zero or their hundred (p.231)

Finally, in order to find out if the limit cycle is the result of a very particular choice of the parameters, it is worth performing a bifurcation analysis of the model with the aim of detecting which are the most relevant parameters. Moreover, it is interesting to check that amplitude and frequency of the cycle are not too sensitive to the parameters because only under this condition is the discovered cycle credible.

Three bifurcation diagrams, obtained with numerical continuation techniques, are shown in Figure 10.4. They point out that the parameter space is partitioned into three subregions, (1), (2), and (3), where the state portraits of the model are like those shown in the top right panels. The curves are Hopf (H) and tangent of limit cycles (TC) bifurcations, that separate the three subregions. Point R corresponds to the reference value of the parameters and falls into region (2).

If we start from point R and gradually decrease the first parameter, namely, the parameter on the horizontal axis, keeping the other constant, *i.e.*, if we move from R toward A', the cycle gradually shrinks and finally becomes a stable equilibrium at point A'. Thus, the transition from region (2) to region (1) at point A' is a supercritical (non-catastrophic) Hopf bifurcation (see Appendix). To put it in words, the periodic romantic regime is smoothly substituted by a stationary regime if the first parameter becomes too small. In contrast, if we start from point R and increase the first parameter, the cycle disappears only when the bifurcation curve TC is crossed at point A'', where the stable and unstable cycles coexisting in region (3) collide (tangent bifurcation of limit cycles, see Appendix). In other words, this time the periodic romantic regime disappears suddenly and is substituted by a stationary romantic regime. The same kind of discontinuous (catastrophic) transitions are obtained if, starting from point R, we vary only the second parameter because at least one of the two extreme points B' and B'' of the region of existence of the cycle are on the tangent bifurcation TC.

The property pointed out in Figure 10.4, namely, that the cycle disappears perturbing the parameters in both directions, basically holds true in all cases. This simply means that the psycho-physical traits of Kathe and Jim should not be too weak or too strong to produce romantic ups and downs.

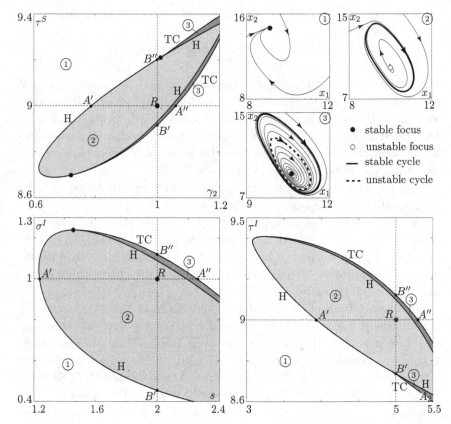

Fig. 10.4   Three bifurcation diagrams of the Kathe-Jim model. For parameters in the three regions ①, ②, and ③ the state portraits are like in the top right panel.

Figure 10.4 shows that there are parameters, namely, Kathe's synergism threshold τ^S and Jim's insecureness threshold τ^I, for which the cycle disappears for quite small perturbations. As done in the previous chapter, parameters for which the cycle is lost by 5% perturbations in both directions are called "critical" in the following. To detect all critical parameters, the bifurcation analysis was systematically performed with respect to all parameters. The result of this systematic analysis is that there are only two critical parameters in the model, namely, those already pointed out in Figure 10.4. A comparison with the previous chapter shows that the Kathe-Jim cycle is definitively more reliable than the Kathe-Jules cycle, and this is possibly due to the fact that Jim's psychological characteristics are not as unusual as those of Jules. Moreover, we have also checked that

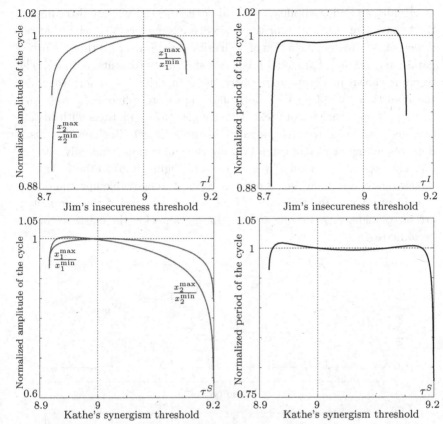

Fig. 10.5   Dependence of the amplitude (first column) and of the period (second column) of the cycle upon the two critical parameters (namely, Kathe's synergism threshold τ^S and Jim's insecureness threshold τ^I). The curves are normalized with respect to amplitude and period of the reference cycle.

the amplitude of the cycle (measured by the ratios $x_i^{\max}/x_i^{\min}$ along the cycle) and its period are not too sensitive to the parameters. This has been done by continuation, the same numerical technique used for bifurcation analysis (see Appendix). The results obtained for the two critical parameters are shown in Figure 10.5, where the curves are normalized to their reference values. The figure shows that the critical parameters have a weak effect on the cycle characteristics in a large portion of the narrow range guaranteeing the existence of the cycle. Thus, we can conclude that almost any value of the critical parameters in the range of existence of the cycle satisfactorily represents the Kathe-Jim love story.

Modeling Love Dynamics

Finally, to add credibility to the discovered cycle, we can also compare the behavior of the model for the reference parameter value with that obtained under 10% perturbation of each critical parameter (in this perturbed parameters setting, the model predicts a stationary romantic regime). The result is shown in Figure 10.6 where the comparison is performed under the realistic assumption that both individuals are influenced, from time to time, by external factors (exogenously generated emotions with instantaneous positive or negative impacts on the feelings). Technically, this is done by adding at random times during the simulation, randomly selected positive or negative amounts (see bottom of Figure 10.6) to the feelings of the two individuals. In such a realistic context, remarkable ups and downs remain present in the time-span of the love story, even if the parameters are set outside, but close to, the region guaranteeing the existence of the periodic regime.

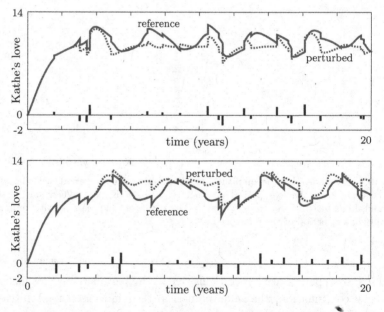

Fig. 10.6 Time evolutions of Kathe's love for reference and 10% perturbed value of each critical parameter (top panel: Kathe's synergism threshold τ^S, bottom panel: Jim's insecureness threshold τ^I). The two time series are obtained by adding to the feelings produced by the model the sequence of random shocks shown at the bottom.

PART II
Complex models

In this part, composed of five chapters, we describe complex models obtained by relaxing the simplifying assumptions we made in the first part. We show that with the addition of one or few variables it is virtually possible to determine the impact of the social environment on romantic relationships, identify the potential consequences of extra emotional dimensions, like artistic inspiration, and extend the analysis to triangular love stories. The analysis from the models suggests a number of interesting conclusions concerning the emergence of turbulence and unpredictability in the evolution of love affairs. These theoretical results are reinforced through the analysis of two famous love stories. The first is described by Francesco Petrarch in his "Canzoniere", the most celebrated collection of love poems in the Western world, and the second is the triangular love story described by Henry-Pierre Roché in his novel "Jules et Jim", popularized by the homonymous film of François Truffaut.

Chapter 11

Environmental stress and romantic chaos

In this chapter we remove one of the main simplifying assumptions made until now, namely, the absence of exogenous events influencing the life of the couple. In the proposed model all the exogenous factors are lumped together and captured by a single variable, called *environmental stress*. Some stresses are strong but cease after a short time (shocks), while others maintain their strength for a long time or have periodic ups and downs. In the model these stresses can be represented with impulse, step, and sinusoidal functions of time. The responses to these particular environmental stresses, called *canonical responses*, are interesting because they mimic a number of real or realistic cases. But they are also interesting because they can be fully derived analytically in the case of linear models (see Chapter 2). This is of great value in all cases in which the stresses are not too heavy because in such cases the couple can be approximately described with a linear model (derived by linearizing the nonlinear model). In the cases of linear or linearized models, the asymptotic patterns of the responses are simply replicates of the patterns of the stresses. For example, if the stress is periodic, the feelings vary in the same way and at the same frequency, in agreement with the theory of linear systems. But if there are nonlinearities and the stresses are quite heavy, the couple can be more complex than a simple replicator. For example, the romantic regime can be chaotic even if the stress is periodic. An example is dedicated to this case and a number of interesting properties are pointed out. In particular, it is shown that *romantic chaos* is more easily promoted if the frequency of the stress (environmental clock) is tuned with that of the couple (romantic clock).

Some knowledge of linear systems theory is certainly helpful for reading the section dedicated to canonical responses, while Lyapunov exponents, Neimark-Sacker, and flip bifurcations are notions that would make the example more transparent.

11.1   Do couples replicate environmental stresses?

Very often romantic relationships are influenced by exogenous factors, here
called environmental stresses. Most of the time, these stresses are unpredictable and cannot be modeled easily. However, at least in principle, we
can imagine that the feelings between two individuals can still be described,
as shown in Figure 11.1, using a model of the same kind we have used until
now, by introducing an extra-variable $w(t)$ representing, in an aggregated
way, all the environmental stresses. Thus, the model is composed of two
ODEs, one for each individual, which can be written in the general form

$$\dot{x}(t) = f(x(t), w(t)). \tag{11.1}$$

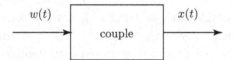

Fig. 11.1   A couple with environmental stress.

Many special, but interesting, cases can be studied with model (11.1). On
one extreme, we have the ideal case in which $w(t)$ can be suitably selected
in order to reach a desired target. For example, one could determine $w(t)$
in order to drive the romantic relationship toward a stationary regime favorable for both partners. A study of this kind can be found in Banerjee
et al. (2015). On the other extreme, we have the case in which $w(t)$ is a
given function of time interpreting the time-varying environmental stress
influencing the couple. For example, the effect of an exogenous shock, *i.e.*, a
very short but important event that suddenly changes the feelings of the
individuals, can be described with a model of the kind

$$\dot{x}(t) = f(x(t)) + cw(t), \tag{11.2}$$

where

$$c = \begin{vmatrix} c_1 \\ c_2 \end{vmatrix}$$

is a two-dimensional vector and $w(t)$ is an *impulse* (see Figure 11.2), *i.e.*, the
function equal to $1/\tau$ in the short time interval $[0, \tau]$ and zero elsewhere
(mathematically speaking, the impulse is obtained in the limit $\tau \to 0$).

In accordance with equation (11.2), the occurrence of an impulse at time $t = 0$, has the effect of varying discontinuously the feeling of individual i from $x_i(0)$ to $x_i(0) + c_i$. Thus, for example, if a couple wins the lottery (Sprott, 2005) the consequences can be studied by determining the response of model (11.2) with $c_1 = c_2$, if the event has the same impact on both individuals. Of course, c_i is negative if the event has a negative impact on individual i. For example, if individual 1 has a secret short extramarital affair of the kind mentioned in Chapter 1, model (11.2) should be used with $c_1 < 0$ and $c_2 = 0$.

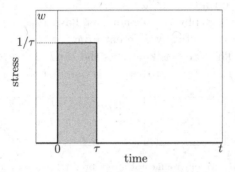

Fig. 11.2 A rectangular impulse of length τ. The mathematical impulse is obtained for $\tau \to 0$.

Other stresses are less intense than the shocks, but persist for some time. For example, a serious physical injury or an operation of aesthetic surgery can modify, negatively or positively, the appeal of an individual for a very long time, if not forever. The consequences of permanent stresses can be detected by using the so-called *step* function $w(t)$ in model (11.2), *i.e.*, the function equal to 0 for $t \le 0$ and to 1 for $t > 0$.

Other interesting stresses are those generated by periodic events impacting on the couple. These are many and are often associated with typical frequencies like the day, the week, the month, and the year. In this case, the function $w(t)$ in model (11.2) should be a *sinusoid* with amplitude W and frequency ω (the corresponding period is $T = 2\pi/\omega$), *i.e.*,

$$w(t) = W \sin \omega t$$

or, more in general, the sum of a number of sinusoids with frequencies ω, 2ω, 3ω, ... because any periodic function of period T can be decomposed in the sum of a sinusoid of frequency $\omega = 2\pi/T$ (first harmonic) and of other sinusoids with frequencies equal to multiples of ω (higher order harmonics). This decomposition of periodic functions is known as a Fourier series.

Finally, in the most complex case, we can use in model (11.2) a function $w(t)$ which is the realization of a suitable stochastic process (see Figure 11.3). The process most often considered by mathematicians is the *white noise*, an extremely abstract stochastic process composed, by definition, of sinusoids of all frequencies with equal and infinitesimal amplitude. White noise is conceptually the mix of a continuum of sinusoids of all possible frequencies, none of which, however, is more relevant than the others. Thus, if a function $S(\omega)$, called *spectrum of the noise*, is a measure of the (infinitesimal) amplitude of the sinusoid with frequency ω, the white noise has a unitary spectrum, *i.e.*, $S(\omega) = 1$. Other noises, called *colored*, used in the study of various physical phenomena, have a spectrum $S(\omega)$ which is not constant and is practically different from zero only in a range $[\underline{\omega}, \bar{\omega}]$ of the frequencies. This range is known as the *band* of the process and when the band is very narrow the process is very similar to a sinusoid.

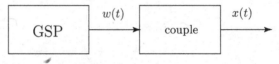

Fig. 11.3   A couple with environmental stress produced by a generator of stochastic processes (GSP).

In all the above cases, the major problem is to detect the characteristics of the feelings in the long term. In particular, one is interested in knowing if the feelings somehow reproduce, at least after some time, the characteristics of the stresses. If this is so, we can say that the couple is simply a sort of *replicator* of the environment. But we already know that this is not always the case. In fact, we have seen in Chapter 3 that in robust couples the effect of a shock vanishes after some time, while in fragile couples, the effect can be permanent (a switch from an unfavorable to a favorable romantic regime). We have also seen that permanent stresses corresponding to parameter variations involving the crossing of a bifurcation curve (for example, a Hopf bifurcation) can involve a radical change in the romantic regime of the couple (for example, from stationary to periodic). Thus, in this case, the environmental stress varies from a constant value to another constant value, while the feelings do not follow a similar pattern because they switch from a stationary to a periodic regime. In the following sections we study the responses to various environmental stresses, with particular attention to the case in which the patterns of the feelings do not replicate those of the stresses. This naturally brings us to investigate a very interesting phenomenon, namely, the emergence of romantic chaos.

11.2 Canonical responses to stress

The stresses $w(t)$ we have mentioned in the previous section, namely, impulses, steps, sinusoids, white and colored noises, are often called "canonical". Consistently, the corresponding feelings, obtained by solving the ODEs (11.1), are called *canonical responses*. These responses are of great interest, not only for the reasons illustrated in the previous section, but also because in the simple case of linear couples they can be derived analytically, thus obtaining valuable theoretical insights.

In line with model (11.2) and Chapter 2, a linear couple with environmental stress is described by

$$\dot{x}_1 = -\alpha_1 x_1 + \beta_1 x_2 + \gamma_1 A_2 + c_1 w$$
$$\dot{x}_2 = -\alpha_2 x_2 + \beta_2 x_1 + \gamma_2 A_1 + c_2 w,$$

or, in more compact form, by

$$\dot{x} = Ax + b + cw, \tag{11.3}$$

where A is a 2x2 matrix and b and c are two-dimensional vectors, *i.e.*,

$$A = \begin{vmatrix} -\alpha_1 & \beta_1 \\ \beta_2 & -\alpha_2 \end{vmatrix} \qquad b = \begin{vmatrix} \gamma_1 A_2 \\ \gamma_2 A_1 \end{vmatrix} \qquad c = \begin{vmatrix} c_1 \\ c_2 \end{vmatrix}.$$

Rewriting equation (11.3) after substituting $\dot{x}$ with sx, we obtain

$$sx = Ax + b + cw, \tag{11.4}$$

where s is a symbol representing the operator performing the time derivative of a function, *i.e.*,

$$s = \frac{d}{dt}$$

Equation (11.4) is a linear algebraic relationship that can be solved with respect to x, giving

$$x = (sI - A)^{-1}b + (sI - A)^{-1}cw, \tag{11.5}$$

where I is the 2x2 identity matrix

$$I = \begin{vmatrix} 1 & 0 \\ 0 & 1 \end{vmatrix}.$$

The first term at the right hand side of equation (11.5) describes the effect of the appeals (appearing in vector b) while the second term is the influence of the environmental stress. Thus, from now on, we only consider the equation

$$x = (sI - A)^{-1}cw \qquad (11.6)$$

since we are interested in evaluating the impact that the environmental stress has on the couple. Since

$$(sI - A)^{-1}c = \begin{vmatrix} s+\alpha_1 & -\beta_1 \\ \\ -\beta_2 & s+\alpha_2 \end{vmatrix}^{-1} \begin{vmatrix} c_1 \\ \\ c_2 \end{vmatrix} = \begin{vmatrix} \dfrac{c_1 s + c_1\alpha_2 + c_2\beta_1}{d(s)} \\ \\ \dfrac{c_2 s + c_2\alpha_1 + c_1\beta_2}{d(s)} \end{vmatrix} = \begin{vmatrix} \dfrac{n_1(s)}{d(s)} \\ \\ \dfrac{n_2(s)}{d(s)} \end{vmatrix}$$

where

$$d(s) = \det(sI - A) = s^2 + (\alpha_1 + \alpha_2)s - \beta_1\beta_2$$

is the characteristic polynomial of A (see Appendix), we can write

$$x_1 = \frac{n_1(s)}{d(s)}w \qquad x_2 = \frac{n_2(s)}{d(s)}w$$

or, equivalently,

$$\begin{aligned} d(s)x_1 &= n_1(s)w \\ d(s)x_2 &= n_2(s)w. \end{aligned} \qquad (11.7)$$

Recalling the meaning of the operator s, the conclusion is that the feelings x_1 and x_2 satisfy the following differential equations of the second order (because $d(s)$ is a polynomial of second degree)

$$\begin{aligned} \ddot{x}_1 + (\alpha_1 + \alpha_2)\dot{x}_1 + (\alpha_1\alpha_2 - \beta_1\beta_2)x_1 &= c_1\dot{w} + (c_1\alpha_2 + c_2\beta_1)w \\ \ddot{x}_2 + (\alpha_1 + \alpha_2)\dot{x}_2 + (\alpha_1\alpha_2 - \beta_1\beta_2)x_2 &= c_2\dot{w} + (c_2\alpha_1 + c_1\beta_2)w. \end{aligned} \qquad (11.8)$$

Thus, the impact of the environmental stress on each individual is the solution of a second order differential equation uniquely identified by the pair of polynomials $(n_i(s), d(s))$. The ratio between these two polynomials

$$G_1(s) = \frac{n_1(s)}{d(s)} \qquad G_2(s) = \frac{n_2(s)}{d(s)}$$

is called *transfer function* and is the most popular tool used for the description of linear systems. In particular, the long-term behavior of the canonical responses can be immediately derived from the transfer functions.

For example, the *step response* is the solution of (11.8) with $w = 1$ and $\dot{w} = 0$, which, after transient, is given by

$$x_1(\infty) = \frac{c_1\alpha_2 + c_2\beta_1}{\alpha_1\alpha_2 - \beta_1\beta_2} = G_1(0) \qquad x_2(\infty) = \frac{c_2\alpha_1 + c_1\beta_2}{\alpha_1\alpha_2 - \beta_1\beta_2} = G_2(0).$$

Thus, in the long term, the feelings replicate the stress since they remain constant forever, and the impact of the stress is equal to the transfer function evaluated for $s = 0$.

The response to a sinusoidal stress

$$w(t) = W \sin \omega t$$

can also be simply determined from the transfer functions. In fact, it is easy to show that the feeling of each individual tends toward a sinusoid with the same frequency ω of the stress, *i.e.*,

$$x_1 = X_1(\omega) \sin(\omega t + \phi_1(\omega)) \qquad x_2 = X_2(\omega) \sin(\omega t + \phi_2(\omega)).$$

The amplitude $X_i(\omega)$ and the phase $\phi_i(\omega)$ of the sinusoid, called *frequency response*, can immediately be computed from the transfer function $G_i(s)$ evaluated for $s = i\omega$ ($i =$ imaginary unit), since

$$\begin{aligned} X_1(\omega) &= |G_1(i\omega)|W & X_2(\omega) &= |G_2(i\omega)|W \\ \phi_1(\omega) &= argG_1(i\omega) & \phi_2(\omega) &= argG_2(i\omega). \end{aligned} \tag{11.9}$$

This proves that the feelings perfectly replicate (with some delay) the sinusoidal pattern of the stress. However, the amplitudes of the sinusoidal responses depend on ω and, actually, tend to 0 for $\omega \to \infty$ because the module of the polynomial $n_i(s)$ evaluated for $s = i\omega$ tends to infinity as ω, while $|d(i\omega)|$ tends to infinity as ω^2. This means that the effect of low frequency environmental stresses can be detected in the feelings, while high frequency periodicities are strongly attenuated at the point of being undetectable. In other words, a couple acts as a low-pass filter.

To illustrate the meaning and the usefulness of frequency response, let us consider a specific couple described, with the month as time unit, by

$$A = \begin{vmatrix} -1 & 0.3 \\ 0.5 & -0.5 \end{vmatrix} \qquad c = \begin{vmatrix} 1 \\ 1 \end{vmatrix}.$$

From the formulas reported above, we obtain

$$G_1(s) = \frac{n_1(s)}{d(s)} = \frac{s + 0.8}{s^2 + 1.5s + 0.35} \qquad G_2(s) = \frac{n_2(s)}{d(s)} = \frac{s + 1.5}{s^2 + 1.5s + 0.35}$$

and

$$X_1(\omega) = |G_1(i\omega)|W = \sqrt{\frac{\omega^2 + 0.64}{\omega^4 + 1.55\omega^2 + 0.1225}}W$$

$$X_2(\omega) = |G_2(i\omega)|W = \sqrt{\frac{\omega^2 + 2.25}{\omega^4 + 1.55\omega^2 + 0.1225}}W.$$

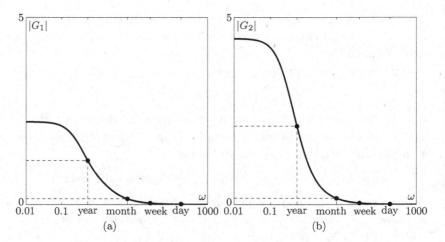

Fig. 11.4 Frequency responses: (a) first individual; (b) second individual. The frequency ω is in logarithmic scale.

The two frequency responses $X_i(\omega)$ are plotted in Figure 11.4 and show that daily and weekly periodicities (corresponding to $\omega = 2\pi/(1/30) = 188.5$ and $\omega = 2\pi/(7/30) = 26.93$) are negligible because their effects on the feelings are practically undetectable, while monthly and yearly periodicities (corresponding to $\omega = 2\pi = 6.28$ and $\omega = 2\pi/12 = 0.52$) must be taken into account because they influence the feelings. All this is consistent with the comments we made in Chapters 1 and 8 when we said that the phenomena a model can claim to describe are those that develop in periods comparable with the time constants ($1/\alpha_1 = 1$ and $1/\alpha_2 = 2$) of the oblivion processes.

If the stress $w(t)$ is a colored noise with spectrum $S(\omega)$, we can apply what we already know for a single sinusoid to each infinitesimal sinusoid composing the noise. The immediate consequence is that the spectrum $S_i(\omega)$ of the feeling x_i is

$$S_i(\omega) = |G_i(i\omega)|S(\omega).$$

In the particular case of white noise, we obtain

$$S_i(\omega) = |G_i(i\omega)|$$

which can be interpreted by saying that if the environment is completely random, in the sense that $w(t)$ is white noise, the spectra of the feelings are the frequency responses of the couple (see (11.9)).

The conclusion of all we have said until now is that linear couples influenced by environmental fluctuations are perfect replicators, in the sense that the feelings of the individuals follow the same pattern as the stresses. Numerical experiments confirming all these conclusions can be found in Wauer *et al.* (2007), where a number of linear models affected by periodic or noisy stresses are studied.

As one can easily imagine, non-linear couples can also be replicators, in particular if the nonlinearities are not too strong. For example, one can expect couples with a unique attractor (an equilibrium or a limit cycle) in the absence of stress to simply replicate the cyclical or noisy patterns of the environmental stresses. This has been ascertained through simulation by Wauer *et al.* (2007), who have studied robust and fragile couples with weak periodic and noisy stresses, and by Barley and Cherif (2011) who have analyzed a quite particular model (one of the two individuals is a "hermit" lover), which was also studied by Ozalp and Koca (2012).

In contrast, other nonlinear couples, like those with alternative stable states, can be expected to respond in a much more subtle way to strong environmental stresses. This is confirmed, for example, by Barley and Cherif (2011) who have shown that a nonlinear model with two alternative stable states reacts to noisy stresses characterized by a unimodal distribution by producing noisy feelings with bimodal distributions. This is because although the feelings remain, most of the time, close to one of the two stable equilibria, they are forced by the noise to switch from time to time to the basin of attraction of the other equilibrium. The fact that even complex responses like these can be intuitively understood and hence predicted from the behavior of couples unaffected by environmental stress is conceptually very important because it reinforces the idea that deterministic models like those we described in the first part of this book can capture even subtle details of love dynamics.

11.3 Romantic chaos

We now show how the problem of chaotic evolution (and hence of unpredictability) of the feelings in couples affected by environmental stress can be discussed by casting that problem in a general conceptual frame.

For this we assume that the dynamics of the environmental stress $w(t)$ are independent of the feelings $x(t)$ and are described by a finite number, say m, of ODEs. Thus, the complete model of the couple has the structure shown in Figure 11.3 and is described by $(m + 2)$ ODEs

$$\dot{w}(t) = g(w(t), p) \tag{11.10}$$

$$\dot{x}(t) = f(x(t), w(t), q) \tag{11.11}$$

where p and q are constant parameters. The Jacobian matrix of the system

$$J = \begin{vmatrix} \dfrac{\partial g}{\partial w} & 0 \\ \dfrac{\partial f}{\partial w} & \dfrac{\partial f}{\partial x} \end{vmatrix}$$

therefore has a triangular structure.

The dynamics of the environmental stress are concisely captured by the so-called Lyapunov exponents (see Appendix), which are m real numbers, that depend on the parameter p which characterizes the attractor (assumed to be unique) of the environmental submodel (11.10). They describe the sensitivity of $w(t)$ to the m components of the initial condition $w(0)$ and can be estimated numerically by suitably averaging the $m \times m$ time-varying matrix $\partial g / \partial w$ along a solution of (11.10) (in the following, we use the algorithm proposed by Ramasubramanian and Sriram (2000)).

Positive Lyapunov exponents reveal the divergence of nearby solutions, typical of chaotic regimes, while negative Lyapunov exponents reveal their convergence. The sign of the largest Lyapunov exponent (LLE) identifies the nature of the attractor: strange attractors (chaotic regimes) have positive LLEs, stable cycles and tori (periodic and quasi-periodic regimes) have LLEs of 0, and stable equilibria (stationary regimes) have negative LLEs. In the following, the LLE of the environmental submodel (11.10) is indicated by L_g and called environmental Lyapunov exponent.

The Lyapunov exponents of the couple, *i.e.*, the Lyapunov exponents of the complete model (11.10,11.11) are therefore $(m + 2)$ real numbers that can be extracted from the Jacobian matrix evaluated along a solution $(w(t), x(t))$ of model (11.10,11.11). In view of the triangular structure of this Jacobian matrix, m Lyapunov exponents of the complete model are those of the environmental submodel (11.10), while the two remaining exponents are generated by the romantic submodel, that is, by the Jacobian $[\partial f / \partial x]$. These two Lyapunov exponents, called romantic, depend not only on q but also on the parameter p because the matrix $[\partial f / \partial x]$ depends on the environmental stress $w(t)$. Thus, the largest romantic Lyapunov exponent, indicated by $L_{f/g}$, is actually conditioned to the characteristics p of the environmental stress. That is why it is called conditional.

From now on, we assume that the environment is chaotic ($L_g > 0$) because the case of constant environment has already been considered in the first part of this book, while environmental periodicities are discussed in the next section, and quasi-periodic environments are of purely mathematical interest.

As for the largest romantic Lyapunov exponent, we consider the following two cases:

$$\text{(a)} \quad L_{f/g} > 0 \qquad \qquad \text{(b)} \quad L_{f/g} < 0,$$

that is, we intentionally rule out from our analysis the particular case $L_{f/g} = 0$, which does not occur generically if the environment is chaotic (this is not so in the next section, where the environment is periodic). In all cases, the couple (*i.e.*, the complete model) is chaotic because its largest Lyapunov exponent $L_{f,g}$ is positive, since

$$L_{f,g} = \max\{L_g, L_{f/g}\} \geq L_g > 0.$$

In case (a), the romantic submodel (11.11) reinforces environmental chaos, while in case (b), the romantic submodel (11.11) does not contribute to the complex behavior of the couple, which is exclusively due to the environment. For this reason we say that in case (a) there is romantic chaos, while in case (b) there is no romantic chaos because the feelings are simply replicates of the environmental stresses. The sign of $L_{f/g}$ is therefore the only information needed for detecting whether chaos in a couple is simply entrained by the environment or is, at least in part, generated by the internal behavioral rules of the couple.

11.4 An example of romantic chaos

In this section we study in some detail a couple that can have romantic chaos for suitable parameter values. To avoid dealing with a completely new case we consider a couple that we have already studied in Chapter 8, where the environmental stress was absent. The couple is composed of two individuals who are insecure and synergic in their reactions to appeal. Thus, the model, in the absence of environmental stress, is

$$\begin{aligned}
\dot{x}_1 &= -\alpha_1 x_1 + R_1^L(x_2) + (1 + b_1^A B_1^A(x_1))\gamma_1 A_2 \\
\dot{x}_2 &= -\alpha_2 x_2 + R_2^L(x_1) + (1 + b_2^A B_2^A(x_2))\gamma_2 A_1,
\end{aligned} \tag{11.12}$$

and the reaction functions $R_i^L(x_j)$ are first increasing and then decreasing (because the individuals are insecure), and the bias coefficients b_i^A are positive (because the individuals are synergic). All functions appearing in model (11.12) are fixed exactly as in Section 8.4 of Chapter 8, *i.e.*,

$$R_1^L(x_2) = \beta_1 k_1 x_2 \exp(-(k_1 x_2)^{n_1}) \quad R_2^L(x_1) = \beta_2 k_2 x_1 \exp(-(k_2 x_1)^{n_2})$$

$$B_1^A(x_1) = x_1^{2m_1}/(x_1^{2m_1} + \sigma_1^{2m_1}) \quad B_2^A(x_2) = x_2^{2m_2}/(x_2^{2m_2} + \sigma_2^{2m_2}).$$

For these functions and reference parameter values

$$\begin{array}{lllll} \alpha_1 = 0.36 & k_1 = 0.08 & n_1 = 1 & \beta_1 = 0.75 & A_1 = 0.1 \\ \sigma_1 = 1 & m_1 = 4 & \gamma_1 = 1 & b_1^A = 2.9 \end{array}$$

$$\begin{array}{lllll} \alpha_2 = 0.2 & k_2 = 1.5 & n_2 = 4 & \beta_2 = 10.66 & A_2 = 0.1 \\ \sigma_2 = 1 & m_2 = 4 & \gamma_2 = 1 & b_2^A = 1. \end{array}$$

model (11.12) has a limit cycle (with a period $T_{cycle} = 19$) as shown in Figure 11.5.

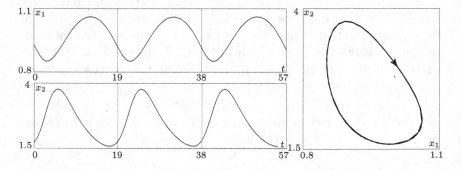

Fig. 11.5 Limit cycle of model (11.12) for the reference parameter values. The period of the cycle is $T_{cycle} = 19$.

Our target is to point out that romantic chaos can emerge when the couple is affected by environmental stress. For this, we consider the case in which the environmental stress is periodic, as this assumption simplifies the analysis. In fact, if the environment varies periodically, *i.e.*, if the attractor of the environmental submodel (11.10) is a limit cycle, the largest environmental Lyapunov exponent L_g is equal to zero, so that $L_{f,g} = L_{f/g}$. In other words, if the environment varies periodically, romantic chaos exists only if the couple is chaotic.

We now proceed by focusing on a particular case (without claiming that it is particularly meaningful). For this, we assume that the appeal A_1

of individual 1 varies sinusoidally over time, peaking, for example, in the period of the summer holidays, *i.e.*,

$$A_1(t) = \bar{A}_1(1 + \varepsilon \sin \omega t) \qquad 0 \leq \varepsilon \leq 1$$

where $\bar{A}_1$ and ε represent the mean value and the relative variation of the appeal of the first individual.

To see if romantic chaos exists, we fix all parameters except two and then perform simulations for a number of combinations of the two free parameters. For example, we fix all parameters of model (11.12) at their reference values and the frequency ω of the environmental stress ($\omega = 2\pi/T$, where $T = 52$ is the length of the year if the adopted time unit is the week) and then simulate the model for various values of the pair ($\varepsilon, \bar{A}_1$). Each simulation must be very long because its initial part (the transient toward the attractor) must be disregarded if we wish to detect if the feelings become, in the long term, periodic, quasi-periodic, or chaotic.

The results of these simulations are long time series of the feelings. Figure 11.6 shows, for example, in its left panels, the evolution of $x_1(t)$ for the six different pairs ($\varepsilon, \bar{A}_1$) indicated in the caption. It is easy to recognize that the first four time series are periodic and that the last two are not, but it is difficult, if not impossible, to say from simple inspection, if they are quasi-periodic or chaotic. Suitable tests must therefore be performed on the time series to discover their nature. A very simple test consists of extracting from each long time series the peaks x_1^h, $h = 1, 2, 3, \ldots$ and in plotting them one versus the previous one. The result is the so-called *peak-to-peak plot* (PPP), which is a visualization of the attractor on a Poincaré section (see Candaten and Rinaldi (2000) and Appendix). Thus, if the PPP contains only one point (or a few points) the time series is periodic because a limit cycle intersects a Poincaré section at a single point (or at a few points). In contrast, if the points of the PPP are densely distributed on a regular closed curve (the section of a torus) the time series is quasi-periodic, while if the PPP is a fractal set the time series is chaotic. The PPPs of the six time series are shown in the right panels of Figure 11.6. The first four allow one to confirm that the corresponding time series are periodic with 1, 2, 5, and 3 peaks per period (note that all periods are equal to 1 year except for the third which is equal to 2 years), while the last two allow one to understand that the corresponding time series are quasi-periodic and chaotic.

To detect the nature of each time series one can complement the simulations with an algorithm for the evaluation of Lyapunov exponents. For

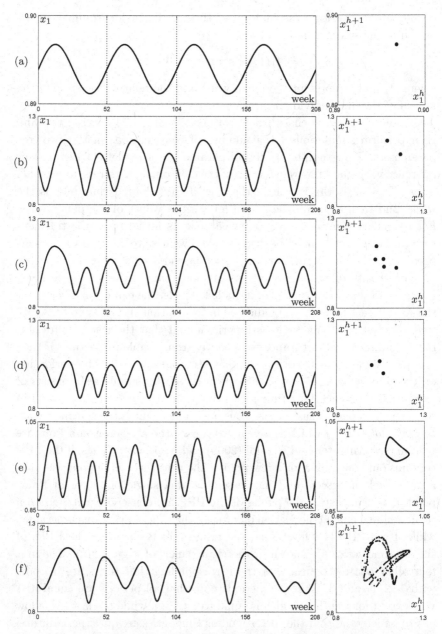

Fig. 11.6   Four years long time series of x_1 obtained through simulation of model (11.12) with reference parameter values and A_1 varying sinusoidally as $\bar{A}_1(1 + \varepsilon \sin \omega t)$: (a) $\bar{A}_1 = 0.017$, $\varepsilon = 0.17$; (b) $\bar{A}_1 = 0.14$, $\varepsilon = 0.17$; (c) $\bar{A}_1 = 0.11$, $\varepsilon = 0.37$; (d) $\bar{A}_1 = 0.09$, $\varepsilon = 0.7$; (e) $\bar{A}_1 = 0.075$, $\varepsilon = 0.5$; (f) $\bar{A}_1 = 0.15$, $\varepsilon = 0.26$. On the right of each time series is the corresponding peak-to-peak plot (PPP).

example, the algorithm of Ramasubramanian and Sriram (2000) gives for the six cases of Figure 11.6 the following estimates of the largest romantic Lyapunov exponent

$$L_{f/g} = -0.11 \qquad L_{f/g} = -0.0074 \qquad L_{f/g} = -0.012 \mid$$
$$L_{f/g} = -0.0042 \qquad L_{f/g} = 0.00005 \qquad L_{f/g} = 0.012.$$

The signs of these six estimates and the fact that the fifth is practically zero reveal that the first four time series are periodic, the fifth is quasi-periodic, and the last is chaotic, as already established from the PPPs of Figure 11.6.

If we wish to discuss the existence of romantic chaos more deeply we must compute the largest romantic Lyapunov exponent more systematically for many parameter settings and extract general messages from the results. A simple way is to fix all parameters except two and evaluate all Lyapunov exponents of the model (and, hence, also $L_{f/g}$) in a large region of the two-dimensional space of the free parameters. Then, this operation can be repeated for other free parameters, or at least for those that are suspected, for some reason, of playing some role in promoting romantic chaos. For example, in the case of our couple of individuals, we may be interested in gaining insights in the role played by mean and variance of the appeal. If so, we must compute the largest romantic Lyapunov exponent $L_{f/g}$ on a dense grid covering a large region of the space $(\varepsilon, \bar{A}_1)$.

Before performing the computations it is worth asking if the theory of dynamical systems can tell us something beforehand. In the present case, we know that on the vertical axis of the space $(\varepsilon, \bar{A}_1)$, *i.e.*, in the absence of environmental stress, the couple has a limit cycle for $\bar{A}_1 = 0.1$ and, presumably, on the basis of the discussion presented in Chapter 8, a Hopf bifurcation for a value A_1^H of the appeal. If, for intuitive reasons, we take as granted that the appeal is a destabilizing factor, we should expect the model unaffected by noise to have a stable equilibrium for

$$\bar{A}_1 < A_1^H \tag{11.13}$$

and a stable cycle when the opposite inequality holds. All this can be confirmed by a series of simulations of the model for different values of $\bar{A}_1$ and $\varepsilon = 0$ or, better, by a bifurcation analysis with respect to $\bar{A}_1$, with $\varepsilon = 0$. In particular, it can easily be established that the cycle exists for $\bar{A}_1 > 0.07$ and that its period depends upon $\bar{A}_1$ as explicitly shown in Figure 11.7, where the points A, B, C, and D are those at which the period of the cycle is equal to 52, 26, 20.8, and 17.3 weeks, that is, 1/1, 1/2, 2/5, and 1/3 times the period of the environmental stress (equal to 52 weeks).

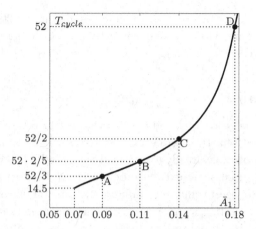

Fig. 11.7 The period of the cycle in the absence of environmental stress increases for $\bar{A}_1 > A_1^H = 0.07$.

This means that the environmental and the romantic clocks beat more and more synchronously when $\bar{A}_1$ is increased.

We can now introduce some small variability in the appeal, *i.e.*, assume that ε is positive but small. If condition (11.13) is satisfied, we are close to the lower part of the vertical axis in Figure 11.8. As, in that region, the couple can be approximated as linear (because ε is small) and is affected by a sinusoidally varying stress, its response should simply be a sinusoidal response. In other words, for sufficiently low values of ε and $\bar{A}_1$ (precisely, $\bar{A}_1 < 0.07$) we expect the couple to have a periodic romantic regime, as qualitatively indicated in Figure 11.8. Similarly, close to the vertical axis above the Hopf bifurcation point $A_1^H = 0.07$, the couple is approximately a periodically varying linear system forced by a sinusoidal input and the expected behavior is quasi-periodic, as sketched in Figure 11.8. In conclusion, for small values of ε the behavior of the couple is periodic if $\bar{A}_1$ is low and quasi-periodic if $\bar{A}_1$ is high. The boundary between these two regions is a Neimark-Sacker bifurcation NS (see Appendix) that can, actually, be produced very precisely with continuation algorithms. But more can be said on the behavior of the couple for ε small, as specified in Figure 11.8 where a few regions (called *Arnold's tongues*) in which the regime is periodic are qualitatively sketched. When ε is very small, these regions are so tiny that they cannot be seen (the distance between their lower and upper boundaries is smaller than a pixel). Each Arnold's tongue is rooted at a point $\bar{A}_1$ on the vertical axis, where for $\varepsilon = 0$ the ratio between the pe-

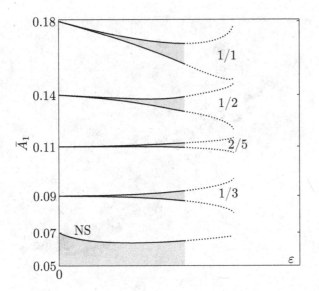

Fig. 11.8 A sketch of the expected romantic regimes for small values of ε: in the gray areas the expected behavior is periodic, while it is quasi-periodic in the white areas above the Neimark-Sacker bifurcation curve NS rooted at point $A_1^H = 0.07$ on the vertical axis.

riod of the cycle (indicated as T_{cycle} in Figure 11.7) and the period of the environmental stress is a rational number, *i.e.*,

$$\frac{T_{cycle}}{T_{env}} = \frac{p}{q}, \tag{11.14}$$

where p and q are integer numbers. For example, Figure 11.7 shows the values of $\bar{A}_1$ for which p/q is equal to 1/1, 1/2, 2/5, and 1/3.

Equation (11.14) shows that in an interval of time equal to $pT_{env}(= qT_{cycle})$ the environmental stress goes through p complete cycles while the couple goes through q romantic cycles. In other words, recalling that $T_{env} = 1$ year, the system (couple and environment) has a p years long periodic behavior characterized by q (identical) peaks of the feelings. For example, if the value of $\bar{A}_1$ corresponds to point C in Figure 11.7, there are 5 romantic peaks every 2 years. If a small seasonality ε is introduced while $\bar{A}_1$ is not varied, one can expect that the p years long cycle with q romantic peaks is not destroyed but that the q romantic peaks do not remain identical. Moreover, it can be shown that this rather special periodic behavior remains alive even if $\bar{A}_1$ is slightly perturbed (from the lower to the upper boundary of the tongue). This phenomenon, often referred to as *frequency locking*, is difficult to point out, not only empirically but also numerically, because

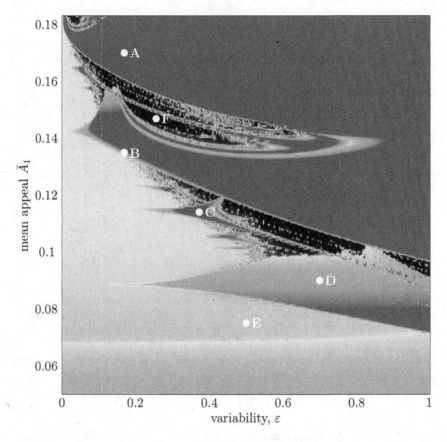

Fig. 11.9   The largest romantic Lyapunov exponent for various values of the variability ε and the mean appeal $\bar{A}_1$. In the green regions below the Neimark-Sacker bifurcation and in the Arnold's tongues the behavior is periodic; in the yellow region the behavior is quasi-periodic; in the red regions the behavior is chaotic.

the p/q tongues are very tiny, in particular if p and q are large.

Now that we have qualitatively seen what the theory suggests, we can look at the precise behavior of the couple by numerically computing the largest romantic Lyapunov exponent $L_{f/g}$ on a dense grid in the space $(\varepsilon, \bar{A}_1)$. The result, shown in Figure 11.9, is in very good agreement with all that we have said so far. First of all, a curve separates the lower green region (where the regime is periodic) from the large yellow region (where the behavior is quasi-periodic). As expected, this curve is the Neimark-Sacker curve emanating from point A_1^H on the vertical axis. Second, a great number of Arnold's tongues, where the behavior is periodic but locked on torus, are clearly visible. The largest of them are rooted at the points of the vertical axis corresponding to points A, B, C, and D in Figure 11.7 where p/q is, respectively, equal to 1/1, 1/2, 2/5, and 1/3. Consistently, in the highest tongue the cycles are one year long and have only one peak per cycle, as clearly indicated in Figure 11.6a corresponding to point A in Figure 11.9. Similarly, in the other three tongues there are q (*i.e.*, 2, 5, and 3) peaks per cycle, as shown in Figure 11.6b, 11.6c, and 11.6d corresponding to the points B, C, and D in Figure 11.9.

But Figure 11.9 also points out red regions where the regime is chaotic (as seen in Figure 11.6f corresponding to point F in Figure 11.9).The existence and the geometric nature (fractal) of these regions are due to phenomena that occur with strong seasonalities (ε large), and this is why they cannot be predicted from the theory concerning low seasonalities (ε small). Some interesting features of these chaotic regions could be discovered through standard bifurcation analysis. For example, one could show that cascades of flip bifurcations are found when approaching these regions in parameter space (Feigenbaum's route to chaos). But we do not expand this issue here, because to do so would bring us too deeply into the mathematical sphere. Instead, we note from Figure 11.9 that the red chaotic regions are regularly distributed in the space $(\varepsilon, \bar{A}_1)$ and that the degree of seasonality ε required to generate chaos is lower for higher values of $\bar{A}_1$. But recalling Figure 11.7, we can also conclude that less environmental variability ε is needed to generate chaos if the romantic and environmental clocks beat at comparable frequencies. This message is perhaps the most valuable conclusion emerging from this relatively complex chapter. Interestingly, a similar principle has also been discovered in biology (Colombo *et al.*, 2008).

Chapter 12

Extra emotional dimensions

Until now, individuals have been assumed to be endowed with a single emotional dimension—the romantic sphere. In reality, almost all individuals have other emotional dimensions some of which interfere with the romantic dimension. For example, involvement in professional activity can strongly influence the quality of romantic relationships.

If the extra emotional dimensions are not influenced by the romantic sphere, we are in the case considered in the previous chapter because the extra emotional dimensions are equivalent to exogenous sources of environmental stress. If, in contrast, the feelings influence and are influenced by the other emotional dimensions, the behavior of the couple can be much more complex since all dimensions are fully involved in the formation of love dynamics. This is the case we study in this and the next chapter.

The existence of multiple dimensions requires the use of models with three or more ODEs, and this means that romantic chaos can generically arise in these couples. Of course, this occurs if the extra emotional dimensions have a destabilizing effect, as in the case examined in the next chapter. But the effect can also be stabilizing, as shown in this chapter, for a hypothetical couple of artists endowed with artistic inspiration. The special case in which inspiration varies much more quickly than feelings is also briefly discussed, while details on the dual case of fast feelings can be found in Rinaldi (1998a).

No new mathematical prerequisites are needed for reading this chapter.

12.1 Introduction

Until now, we have studied couples composed of individuals with a single emotional dimension, called the romantic dimension (sphere) and identified by a single variable, the feeling for the partner. In reality, almost all individuals have many other emotional dimensions, some of which influence and are influenced by the romantic sphere. Poets are typical examples of individuals with multiple emotional dimensions. Often they recharge their inspiration when they are in love, but at the same time they lose interest in the appeal of the partner when they are highly inspired. The same behavior holds true not only for all kinds of artists but also for scientists, politicians, managers, and, more in general, for people involved in creative professions. The models used until now are composed of two ODEs, one for "she" and one for "he", written in the form

$$\dot{x} = f(x), \tag{12.1}$$

where $x = (x_1, x_2)$ are the variables fully describing the two romantic dimensions. If each individual has other emotional dimensions, other variables, say $z_1^{(1)}$, $z_1^{(2)}$, ... for individual 1 and $z_2^{(1)}$, $z_2^{(2)}$, ... for individual 2, must be introduced to capture all emotional involvements. To avoid heavy notations, we suppose from now on that each individual has only one extra emotional dimension. Thus, each individual i is identified by a pair of variables (x_i, z_i), where x_i is the feeling for the partner and z_i is the extra emotional involvement. If we assume that all dimensions interact, the model should therefore be of the general form

$$\begin{aligned} \dot{x} &= f(x, z) \\ \dot{z} &= g(x, z). \end{aligned} \tag{12.2}$$

If the extra emotional dimensions are not influenced by love, the function g depends only on z and the model becomes

$$\begin{aligned} \dot{x} &= f(x, z) \\ \dot{z} &= g(z), \end{aligned}$$

which is a special case of the model discussed in the previous chapter. Thus, in order to deal with a new case, we assume that the function g in (12.2) depends on x. However, to keep the complexity of the analysis under control, we consider only the special case in which the extra emotional dimension is influenced and influences only the romantic dimension of the same individual. This means that the model has the influence graph shown in Figure 12.1 and is described by the following four ODEs

$$\begin{aligned} \dot{x}_i &= f_i(x_i, x_j, z_i) \ i = 1, 2 \\ \dot{z}_i &= g_i(x_i, z_i) \quad i = 1, 2. \end{aligned} \tag{12.3}$$

The limitation to the special model (12.3) is not very restrictive because interpersonal influence among extra emotional dimensions is certainly rare.

Fig. 12.1 Influence graph of a couple composed of two individuals with one extra emotional dimension.

12.2 The model

Consider a couple composed of two artists who are insecure and synergic in their reaction to appeal but attenuate their reactions to love and appeal when they are inspired (in a sense they become platonic when they are inspired). Assume also that inspiration decays to zero if it is not sustained by love. Thus, the model is

$$
\begin{aligned}
\dot{x}_1 &= -\alpha_1 x_1 + (R_1^L(x_2) + (1 + b_1^A B_1^A(x_1))\gamma_1 A_2)\frac{1}{1 + \delta_1 z_1} \\
\dot{x}_2 &= -\alpha_2 x_2 + (R_2^L(x_1) + (1 + b_2^A B_2^A(x_2))\gamma_2 A_1)\frac{1}{1 + \delta_2 z_2} \\
\dot{z}_1 &= \varepsilon_1(\mu_1 x_1 - z_1) \\
\dot{z}_2 &= \varepsilon_2(\mu_2 x_2 - z_2),
\end{aligned} \tag{12.4}
$$

where the reactions $R_i^L(x_j)$ are first increasing and then decreasing and the two bias coefficients b_i^A are positive because the two artists are insecure and synergic. The parameter δ_i is a measure of the inhibition that the inspiration z_i has on the reactions to love and appeal of individual i. The parameters ε_i and μ_i identify the dynamics of inspiration. More precisely, the inverse of ε_i is the time constant T_i at which inspiration exponentially decays to zero if not sustained by love ($x_i = 0$ implies $\dot{z}_i = -\varepsilon_i z_i$, that is, $z_i(t) = z_i(0) \exp(-t/T_i)$) while μ_i is the conversion factor of love into inspiration (at equilibrium $z_i = \mu_i x_i$).

If there is no inhibition, i.e., if $\delta_1 = \delta_2 = 0$, the first two ODEs coincide with those of the model discussed in Section 8.4. Thus, we already know that the two artists can have romantic ups and downs if their reactions are not inhibited by inspiration.

12.3   Analysis and results

To simplify the analysis, we select the functions $R_i^L(x_j)$ and $B_i^A(x_i)$ exactly as in Chapter 8, *i.e.*,

$$R_1^L(x_2) = \beta_1 k_1 x_2 \exp(-(k_1 x_2)^{n_1}) \quad R_2^L(x_1) = \beta_2 k_2 x_1 \exp(-(k_2 x_1)^{n_2})$$

$$B_1^A(x_1) = x_1^{2m_1}/(x_1^{2m_1} + \sigma_1^{2m_1}) \quad B_2^A(x_2) = x_2^{2m_2}/(x_2^{2m_2} + \sigma_2^{2m_2}),$$

and we fix the parameters at the same reference values, *i.e.*,

$$\begin{array}{llllll}
\alpha_1 = 0.36 & k_1 = 0.08 & n_1 = 1 & \beta_1 = 0.75 & A_1 = 0.1 & \sigma_1 = 1 \\
m_1 = 4 & \gamma_1 = 1 & b_1^A = 2.9 & \mu_1 = 0.1 & \varepsilon_1 = 0.1
\end{array}$$

$$\begin{array}{llllll}
\alpha_2 = 0.2 & k_2 = 1.5 & n_2 = 4 & \beta_2 = 10.66 & A_2 = 0.1 & \sigma_2 = 1 \\
m_2 = 4 & \gamma_2 = 1 & b_2^A = 1 & \mu_2 = 0.1 & \varepsilon_2 = 0.1.
\end{array}$$

Note that the two inhibition parameters δ_1 and δ_2 are not specified, at least for the moment. As already noted, if $\delta_1 = \delta_2 = 0$, the two artists have periodic ups and downs as shown in Figure 11.5 in Chapter 11 and also reported (labeled with 0) in Figure 12.2b. Thus, model (12.4) for $\delta_1 = \delta_2 = 0$ has a limit cycle in the space (x_1, x_2, z_1, z_2). We can therefore, with a suitable algorithm, continue this limit cycle along any line starting from the origin in the parameter space (δ_1, δ_2), for example, along the line $\delta_1 = \delta_2$, as shown in Figure 12.2a. More simply, starting from any point $(x_1(0), x_2(0), z_1(0), z_2(0))$ of the cycle with $\delta_1 = \delta_2 = 0$, we can

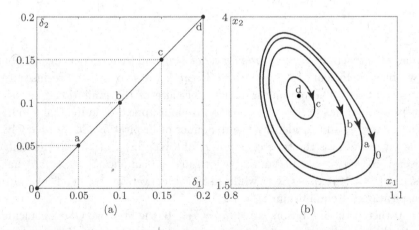

Fig. 12.2   (a) Line $\delta_1 = \delta_2$ in parameter space along which simulations are performed. (b) Projections of the limit cycles in the space of the feelings for the values δ_i indicated in (a).

perform a first simulation of model (12.4) for $\delta_1 = \delta_2 = \delta$, where δ is small. The result of this simulation is a slightly different limit cycle in the four-dimensional space that can be used, again, to produce new initial conditions $(x_1(0), x_2(0), z_1(0), z_2(0))$ for a second simulation of the model for a slightly bigger value of δ. And we can continue in the same way by incrementing δ step by step along the line shown in Figure 12.2a. The result is a sequence of limit cycles, the projections of which in the two-dimensional space (x_1, x_2) are shown in Figure 12.2b. The fact that these cycles shrink with δ and actually disappear for $\delta = 0.157$ is a clear indication of the existence of a Hopf bifurcation for that value of δ.

The Hopf bifurcation can be detected more formally by computing the four eigenvalues of the Jacobian matrix evaluated at the equilibrium for $\delta_1 = \delta_2 = \delta$, which, after a bit of algebra, can be written in the form

$$
J = \begin{vmatrix}
-\alpha_1 + \frac{\gamma_1 A_2}{1+\delta z_1}\frac{dB_1^A}{dx_1} & \frac{dR_1^L}{dx_2}\frac{1}{1+\delta z_1} & \frac{\alpha_1 \delta x_1}{1+\delta z_1} & 0 \\
\frac{dR_2^L}{dx_1}\frac{1}{1+\delta z_2} & -\alpha_2 + \frac{\gamma_2 A_1}{1+\delta z_2}\frac{dB_2^A}{dx_2} & 0 & \frac{\alpha_2 \delta x_2}{1+\delta z_2} \\
\varepsilon_1 \mu_1 & 0 & -\varepsilon_1 & 0 \\
0 & \varepsilon_2 \mu_2 & 0 & -\varepsilon_2
\end{vmatrix}. \tag{12.5}
$$

The locus showing how the two dominant eigenvalues of J (*i.e.*, those with greatest real part) vary with δ in the complex plane is reported in Figure 12.3. It points out that for $\delta = 0$ the equilibrium is unstable (because for $\delta = 0$ the Jacobian matrix has two eigenvalues with positive real part) and that it remains unstable until $\delta = 0.157$, when the two complex eigen-

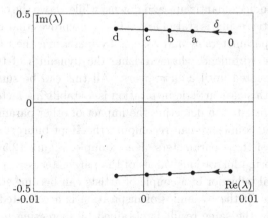

Fig. 12.3 Locus of the two dominant eigenvalues of the Jacobian matrix (12.5), when $\delta = \delta_1 = \delta_2$ is varied. The points 0, a, b, c, d correspond to the values of δ reported in Figure 12.2a.

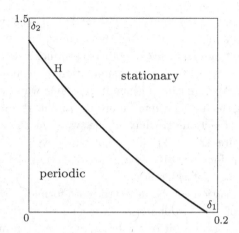

Fig. 12.4   The Hopf bifurcation curve separating stationary from periodic behavior of the couple in the space of the two inhibition parameters.

values cross the imaginary axis. For $\delta > 0.157$ all eigenvalues have negative real parts and the equilibrium is the unique attractor of the system.

Now that we have established that the system has a Hopf bifurcation for $\delta_1 = \delta_2 = 0.157$, we can use a continuation algorithm to produce the entire Hopf bifurcation curve in the space (δ_1, δ_2). The result is presented in Figure 12.4 which shows that couples corresponding to points below the curve have periodic romantic ups and downs, while above the curve the only possible romantic regime is stationary. This figure proves that by increasing the inhibition parameters δ_1 and δ_2, i.e., by increasing the role played by inspiration, the turbulence characterizing the dynamics of the couple is gradually attenuated until it disappears. All this can be summarized by saying that in this case artistic inspiration is a stabilizing factor.

If we are interested in detecting the impact of other parameters on the dynamics of the couple we can recompute the Hopf bifurcation curve for various values of these parameters. For example, Figure 12.5a, where the curve H is reported for various values of the parameter $\varepsilon = \varepsilon_1 = \varepsilon_2$, shows that the cyclical behavior of a couple of artists can become stationary if ε is increased, i.e., if the dynamics of inspiration is accelerated. As shown in Figure 12.5b, the same result is obtained by increasing $\mu = \mu_1 = \mu_2$, i.e., by increasing the conversion factor of love into inspiration. All these results confirm that artistic inspiration is a stabilizing factor.

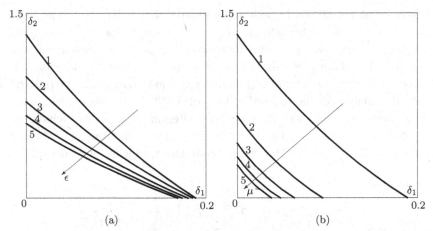

Fig. 12.5 The dependence of the Hopf bifurcation curves on the two extra parameters ε and μ identifying the dynamics of the extra emotional dimensions: (a) ε =speed at which inspiration varies; (b) μ =conversion factor of love into inspiration. The curves show that artistic inspiration is a stabilizing factor. The numbers reported on each curve are 10ϵ and 10μ, respectively.

12.4 Slow-fast dynamics

The results we presented in the previous section were obtained through numerical bifurcation analysis. In general, the same results cannot be obtained through simpler approaches unless special conditions suggest the use of approximated models. Quite interesting, in this sense, is the case in which the variables of the model can be partitioned into two groups: slow and fast variables. In the present case this can occur in two ways. Either inspirations vary more quickly than feelings, or the feelings are the fast variables. To be in the fast inspiration case, the absolute value $|\dot{z}_i|$ of the derivatives of the inspirations must be generically high with respect to the derivatives of the feelings, and this occurs if the two parameters ε_i are large. If this is so, the inspirations z_i tend quickly toward $\mu_i x_i$ so that we can imagine that the slow dynamics of the feelings are well described by the equations

$$\dot{x}_1 = -\alpha_1 x_1 + (R_1^L(x_2) + (1 + b_1^A B_1^A(x_1))\gamma_1 A_2)\frac{1}{1 + \delta_1\mu_1 x_1}$$

$$\dot{x}_2 = -\alpha_2 x_2 + (R_2^L(x_1) + (1 + b_2^A B_2^A(x_2))\gamma_2 A_1)\frac{1}{1 + \delta_2\mu_2 x_2},$$

$$(12.6)$$

obtained by substituting z_i with $\mu_i x_i$ in the first two equations (12.4). Thus, model (12.6) is a reduced-order model of the feelings which, intuitively speaking, should be well approximated if ε_i is large. Proving that this statement is mathematically correct requires the use of the so-called *singular perturbation approach* (Hoppensteadt, 1974). As this is absolutely non-trivial (and not particularly interesting for the non-mathematically oriented readers) we here avoid getting into that kind of detail and we proceed with the analysis of the reduced-order model (12.6). As this model is two-dimensional, we can check if Bendixon's criterion allows us to exclude the existence of limit cycles, *i.e.*, the possibility of periodic romantic ups and downs in the couple. For this, we compute the divergence of model (12.6)

$$\text{div} f = -\alpha_1 - \alpha_2 +$$
$$- (R_1^L(x_2) + (1 + b_1^A B_1^A(x_1))\gamma_1 A_2)\frac{\delta_1 \mu_1}{(1 + \delta_1 \mu_1 x_1)^2} +$$
$$- (R_2^L(x_1) + (1 + b_2^A B_2^A(x_2))\gamma_2 A_1)\frac{\delta_2 \mu_2}{(1 + \delta_2 \mu_2 x_2)^2} +$$
$$+ \gamma_1 A_2 b_1^A \frac{dB_1^A}{dx_1} + \gamma_2 A_1 b_2^A \frac{dB_2^A}{dx_2}$$

and we note that the first four terms are negative while the last two are positive, so that nothing can be said, unfortunately, about the sign of the divergence. However, for high values of μ_i, *i.e.*, for artists strongly inspired by love, the third and fourth terms are negligible so that the sign of the divergence can be easily ascertained by comparing the forgetting coefficients α_i with the terms $\gamma_i A_j b_i^A \, dB_i^A/dx_i$. In particular, if the appeals A_i of the individuals are both very low [high] the divergence is negative [positive] so that Bendixon's criterion excludes the existence of limit cycles. In conclusion, under these assumptions, artists with high ε_i and μ_i, *i.e.*, artists quickly and strongly inspired by love, cannot have romantic ups and downs. This conclusion, obtained through relatively simple algebraic manipulations, is consistent with the results shown in Figure 12.5 derived through numerical bifurcation analysis.

Of course, the singular perturbation approach can also be used to study the dual case (ε_i small), in which inspirations vary much more slowly than feelings. This is not done here because an example of this kind can be found in Rinaldi (1998a), where the conclusion is the opposite of the one we have just pointed out, namely, inspiration can be a destabilizing factor.

Chapter 13

Laura and Francesco

In this chapter we study the 21-year love story between Laura, a beautiful but married lady, and Francesco Petrarch a famous Italian poet of the XIV century (see Figure 13.1). Their relationship is described in the "Canzoniere," one of the most celebrated collection of love poems in the Western world. She is strongly insecure; he is secure but with a reaction to appeal inhibited by his poetic inspiration, described by the equation used in the previous chapter. Thus, only one of the two individuals has an extra emotional dimension and the model is three-dimensional.

Laura de Sade   Francesco Petrarch

Fig. 13.1 Portraits of Laura and Petrarch (Biblioteca Medicea Laurenziana, Firenze, Italy; courtesy of Ministero per i Beni Culturali e Ambientali).

The calibration of the model is described in some detail taking into

149

account the study of Jones (1995), who has obtained, through a careful linguistic and lyrical analysis, a quantitative estimate of the involvement of the poet at 23 particular dates in the period of concern. The model shows that the love story quickly evolves toward a romantic cycle which is in very good agreement with the 23 estimates of Jones.

A very important byproduct of this study is that it strongly reinforces the popular belief that Petrarch was responsible for the spectacular cultural transition from the Middle Ages to Humanism.

No new mathematical prerequisites are needed for reading this chapter. More details on the analysis of the model can be found in Rinaldi (1998a), where the case of slow inspiration and fast feelings is also studied through the singular perturbation approach. A description of the love story as an optimal control problem can be found in Feichtinger *et al.* (1999), while an alternative approach, based on viability theory, is proposed in Bonneuil (2014), where the romantic cycle is interpreted in terms of emotional fluctuations. Finally, it is also worth mentioning that the Laura and Francesco model has been used in numerous courses and conferences to point out the power of the modeling approach in the study of literature, poetry, and film (Breitenecker *et al.*, 2008; Koss, 2015; Zhuravlev *et al.*, 2014).

13.1   Petrarch's romantic cycle

Francesco Petrarch (1304-1374), arguably the most lovesick poet of all time, is the author of the "Canzoniere," a collection of 366 poems (sonnets, songs, sestinas, ballads, and madrigals). In Avignon, at the age of 23, he met Laura, a beautiful but married lady. He immediately fell in love with her and, although his love was not reciprocated, he addressed more than 200 poems to her over the next 21 years. The poems express bouts of ardor and despair, snubs and reconciliations, and they mark the birth of modern love poetry. The verse has influenced countless poets, including Shakespeare.

Unfortunately, only a few lyrics of the "Canzoniere" have an actual date; the rest are collected in a baffingly obscure order. The knowledge of the correct chronological order of the poems is a prerequisite for studying the lyrical, psychological, and stylistic development of any poet. This fact is particularly relevant for Petrarch, as will be shortly discussed in the last section of the chapter. For this reason, the identification of the chronological order of the poems of the "Canzoniere" has been a problem of major concern for centuries.

Frederic Jones (1995) has described how he solved the ordering problem. First, he noticed that in a number of lyrics Petrarch makes reference to the recurrent nature of his amorous experience. For example, in sonnet LXXVI he says (here and in the following quotations the English version is taken from an English translation of the "Canzoniere" made available by Frederic Jones)

Amor con sue promesse lusingando	*Love's promises so softly flattering me*
mi ricondusse alla prigione antica	*have led me back to my old prison's thrall*

while in sonnet CCXXI he asks

Quale mio destin,	*What fate,*
qual forza o qual inganno	*what power or what insidiousness*
mi riconduce	*still guides me back,*
disarmato al campo	*disarmed, to that same field*
là 've sempre son vinto?	*wherein I'm always crushed?*

Laura's attitude is also reported to have periodically softened. For example, in ballad CXLIX Petrarch says

Di tempo in tempo mi si fa men dura	*From time to time less reproachful seem to me*
l'angelica figura e'l dolce riso,	*her heavenly figure, and her charming face,*
et l'aria del bel viso	*and sweet smile's airy grace,*
e degli occhi leggiadri men oscura	*while her dancing eyes grow far less dark I see*

Second, Jones collected all dated poems written when Laura was alive. These amount to 42, but only 23 have a fairly secure date. The first (sonnet X) was written in 1330 and the last (sonnet CCXII) in 1347. Jones thoroughly analyzed each of these 23 poems from a linguistic and lyrical point of view. On the basis of this analysis (described in detail in the third chapter of his book) he assigned a grade ranging from -1 to +1 to each poem. The maximum grade (+1) stands for ecstatic love, while very negative grades correspond to deep despair, as in sonnet LXXIX, where Petrarch says

Così mancando vo di giorno in giorno,	*Therefore my strength is ebbing day by day,*
sì chiusamente, ch'i' sol me ne accorgo	*Which I alone can secretly survey,*
et quella che guardando il cor mi strugge	*and she whose very glance*
	will scourge my heart.

Intermediate grades indicate less extreme feelings like ardor, serene love, friendship, melancholy, and anguish. For example, sonnet CLXXVI, where Petrarch says,

Parme d'udirla, udendo i rami et l'ore   Her I seem to hear, hearing
                                                      bough and wind's caress,
et le frondi, et gli augei lagnarsi, et l'acque   as birds and leaves lament,
                                                      as murmuring flees
mormorando fuggir per l'erba verde.   the streamlet coursing
                                                      through the grasses green.

is graded -0.45 (corresponding to melancholy).

The dates and the grades given by Jones to the 23 poems are reported as points in Figure 13.2, together with four rising segments of a dotted line interpolating some of the points. The distances between pairs of nearby segments are not very different, so that one is naturally led to imagine that these segments are "fragments" of a cyclical pattern. In other words, the analysis indicates that Francesco's feelings for Laura have varied periodically over time and that the period of this romantic cycle is slightly less than four years.

This discovery has been fully exploited by Jones. Indeed, he has extrapolated using basic techniques the segments of Figure 13.2, thus deriving an almost cyclical graph $x_2(t)$ describing the time pattern of Francesco's feelings during the entire interval concerned [1330, 1348]. Then, he has given a grade $\hat{x}_2$ to each undated poem and derived a set $\{t_i\}$ of potential dates for each sonnet by setting $x_2(t_i) = \hat{x}_2$. Finally, using historical and other

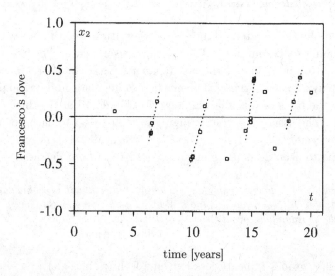

Fig. 13.2 The coordinates of the points are the dates and the grades of the 23 poems analyzed by Jones (1995). The dotted lines are fragments of Francesco's romantic cycle.

information about Petrarch's life and his visits to Avignon, he has eliminated all but one potential date for each poem, thus solving the ordering problem completely.

The key point of the study is undoubtedly the discovery of the Laura and Francesco (L-F) romantic cycle. But, in a sense, this discovery is only supported by relatively weak arguments. Indeed, it is based on noisy data, because each point of Figure 13.2 is affected by a horizontal error, due to the uncertainty of the date of the poem, and by a vertical error due to Jones's subjective evaluation of the poet's involvement. Independent arguments supporting the same conclusion are therefore important.

13.2   A model of Laura and Francesco

The feelings of Laura and Francesco are now modeled by means of three ODEs. Laura is described by a single variable $x_1(t)$, representing her love for the poet at time t. Positive and high values of x_1 mean warm friendship, while negative values are associated with coldness and antagonism. The personality of the poet is more complex; its description requires two variables: $x_2(t)$, love for Laura, and $z_2(t)$, poetic inspiration. High values of x_2 indicate ecstatic love, while negative values stand for despair.

The Laura-Francesco model (from now on called L-F model) is as follows:

$$\dot{x}_1 = -\alpha_1 x_1 + R_1^L(x_2) + \gamma_1 A_2 \tag{13.1}$$

$$\dot{x}_2 = -\alpha_2 x_2 + R_2^L(x_1) + \gamma_2 A_1 \frac{1}{1 + \delta z_2} \tag{13.2}$$

$$\dot{z}_2 = \varepsilon(\mu x_2 - z_2), \tag{13.3}$$

where $R_1^L(x_2)$ and $R_2^L(x_1)$ are reactions to love, A_2 [A_1] is the appeal of Petrarch [Laura], and all Greek letters are positive constant parameters (this means that variations in the personalities of Laura and Francesco due to aging or other external factors are not considered). Equation (13.3) takes into account the fact that high moral tensions, associated with artistic inspiration, attenuate the role of the most basic instincts. And there is no doubt that the tensions between Petrarch and Laura are of a passionate nature. For example, in sestina XXII, he says

Con lei foss'io da che si parte il sole,   Would I were with her when first sets the sun,
et non ci vedess' altri che le stelle,   and no one else could see us but the stars,
sol una nocte, et mai non fosse l'alba   one night alone, and it were never dawn

while in his "Posteritati" he confesses (in Latin): *"Libidem me prorsus ex-pertem dicere posse optarem quidem, sed si dicat mentiar"* [I would truly like to say absolutely that I was without libidinousness, but if I said so I would be lying]. Finally, (13.3) simply says that the love of the poet sustains his inspiration which, otherwise, would exponentially decay with a time constant $1/\varepsilon$. In other words, poetic inspiration is an exponentially weighted integral of the passion of the poet for his mistress.

The two reaction functions $R_1^L(x_2)$ and $R_2^L(x_1)$ must now be specified. The most simplistic choice would be to assume that the reaction functions are linear. The linearity of $R_2^L(x_1)$ is undoubtedly acceptable (at least for $x_1 < 0$) as in his poems the poet has very intense reactions to the most relevant signs of antagonism from Laura. Thus, we assume

$$R_2^L(x_1) = \beta_2 x_1. \tag{13.4}$$

In contrast, a linear reaction function is not appropriate for Laura. In fact, only close to the origin can $R_1^L(x_2)$ be assumed to be linear, thus interpreting the natural inclination of a beautiful high-society lady to encourage harmless flirtations. But Laura never goes too far beyond gestures of pure courtesy: she smiles and glances. However, when Francesco becomes more demanding and puts pressure on her, even indirectly when his poems are sung in public, she reacts very promptly and rebuffs him, as described explicitly in a number of poems, like sonnet XXI:

Mille fiate, o dolce mia guerrera,	*A thousand times, o my sweet enemy,*
per aver co' begli occhi vostri pace	*to come to terms with your enchanting eyes*
v'aggio proferto il cor; mà voi non piace	*I've offered you my heart, yet you despise*
mirar sì basso colla mente altera	*aiming so low with mind both proud and free*

This suggests the use of a reaction to love $R_1^L(x_2)$ which, for $x_2 > 0$, first increases and then decreases. But the behavior of Laura is also nonlinear for negative values of x_2. In fact, when $x_2 \ll 0$, *i.e.*, when the poet despairs, Laura feels very sorry for him. Following her genuine Catholic ethic she arrives at the point of overcoming her antagonism by strong feelings of pity, thus reversing her reaction to the passion of the poet. This behavioral characteristic of Laura is repeatedly described in the "Canzoniere". For example, in sonnet LXIII the poet says

Volgendo gli occhi al mio novo colore	*Casting your eyes upon my pallor new,*
che fa di morte rimembrar la gente,	*which thoughts of death recalls to all mankind,*
pietà vi mosse; onde, benignamente	*pity in you I've stirred; whence, by your kind*
salutando, teneste in vita il core.	*greetings, my heart to life's kept true.*

The above is equivalent to saying that the function $R_1^L(x_2)$, besides having a positive maximum for $x_2 > 0$, has a negative minimum for $x_2 < 0$.

In the following, Laura's reaction function $R_1^L(x_2)$ is assumed to be a cubic function, *i.e.*,

$$R_1^L(x_2) = \beta_1 x_2 (1 - (x_2/x_2^*)^2). \tag{13.5}$$

Thus, for $x_2 = x_2^*$ flattery compensates for antagonism (so that $R_1^L(x_2^*) = 0$), while for $x_2 = -x_2^*$ antagonism is compensated for by pity. Obviously, in no way can one support, from the "Canzoniere" and its related historical information, the specific choice (13.5) for Laura's reaction to love. It is better to confess that this choice is due to convenience, as it allows a number of interesting results to be analytically derived, as shown in Rinaldi (1998a).

In conclusion, if one takes into account (13.4) and (13.5), the L-F model (13.1-13.3) becomes

$$\dot{x}_1 = -\alpha_1 x_1 + \beta_1 x_2 (1 - (x_2/x_2^*)^2) + \gamma_1 A_2 \tag{13.6}$$

$$\dot{x}_2 = -\alpha_2 x_2 + \beta_2 x_1 + \gamma_2 A_1 \frac{1}{1 + \delta z_2} \tag{13.7}$$

$$\dot{z}_2 = \varepsilon(\mu x_2 - z_2). \tag{13.8}$$

It is important to note that this model encapsulates the personalities of Laura and Francesco as they emerge from all the poems of the "Canzoniere", and not from the particular 23 poems analyzed by Jones. Thus, the model and the conclusions one can draw from it are "independent" from those obtained by Jones.

13.3   Calibration of the model

The calibration of the parameters of the L-F model is particularly difficult and highly subjective: all available information comes from a series of poems. Therefore the values given to the parameters in this section do not claim to be the "correct" ones. They are only subjective estimates based on the impressions one has when reading the "Canzoniere".

Let us start with the parameters α_i, $i = 1, 2$, that describe the forgetting processes. There is no doubt that

$$\alpha_1 > \alpha_2,$$

as Laura never appears to be strongly involved, while the poet definitely has a tenacious attachment. This is described in a number of lyrics, as in sonnet XXXV:

Solo et pensoso i più deserti campi	*Alone and lost in thought, each lonely strand*
vo mesurando a passi tardi e lenti,	*I measure out with slow and laggard step,*
. . .	
Ma pur sì aspre vie ne' sì selvagge	*Yet I cannot find such harsh and savage trails*
cercar non so ch'Amor non venga sempre	*where love does not pursue me as I go,*
ragionando con meco, et io co llui.	*with me communing, as with him do I.*

On the other hand,

$$\alpha_2 > \varepsilon$$

as the inspiration of the poet wanes very slowly. Indeed, he continues to write (over one hundred poems) for more than ten years after the death of Laura. The main theme of these lyrics is not his passion for Laura, which has long since faded, but the memory for her and the invocation of death. This is clear, for example, in song CCLXVIII, written about two years after her demise, when he says,

Tempo è ben di morire,	*It's time indeed to die,*
et ò tardato più ch'i non vorrei.	*and I have lingered more than I desire.*
Madonna è morta, et à seco il mio core;	*My lady's dead, and with her my heart lies;*
e volendol seguire,	*and, keen with her to fly,*
interromper conven quest'anni rei,	*I now would from this wicked world retire,*
perché mai veder lei	*since I can no more aspire*
di qua non spero, et l'aspettar m'è noia.	*on earth to see her, and delay will me destroy.*

We will consistently constrain α_1, α_2, and ε to satisfy the relationships

$$\alpha_1 = 3\alpha_2 \qquad \varepsilon = \frac{1}{10}\alpha_2. \qquad (13.9)$$

For example, the triplet

$$\alpha_1 = 3, \qquad \alpha_2 = 1, \qquad \varepsilon = 0.1$$

satisfies (13.9) and can be interpreted by imagining that Laura forgets Francesco in about four months and his passion fades in one year, whereas he remains inspired for ten years.

As far as the reactiveness to love β_i are concerned, again Laura is much less sensitive than Francesco ($\beta_1 < \beta_2$) and the poet is very strongly inspired by love. For these reasons we fix

$$\beta_1 = \alpha_2, \qquad \beta_2 = 5\alpha_2, \qquad (13.10)$$

which is equivalent to saying that he is five times more reactive than her to love.

Moreover, we assume that

$$\gamma_1 = \beta_1, \qquad \gamma_2 = \beta_2, \qquad x_2^* = \delta = 1, \tag{13.11}$$

as this is always possible by suitably scaling x_2 and z_2, and we fix

$$\mu = 100 \tag{13.12}$$

as the poet is very strongly inspired by love.

Finally, opposite signs are given to the appeals of Laura and Francesco (see Figure 13.1), namely,

$$A_1 = 2, \qquad A_2 = -1. \tag{13.13}$$

Indeed, as repeatedly described in the "Canzoniere," she is a beautiful and inspiring lady, while he is just the opposite as recognized by the poet himself, who, in sonnet XLV, while talking about Laura's mirror, says

Il mio adversario *in cui veder solete*	My rival *in whose depths you're wont to see*
gli occhi vostri ch'Amore e 'l ciel honora	*your own dear eyes which Love*
	and heaven apprize

Once Francesco's forgetting coefficient α_2 is fixed, (13.10-13.13) produce one complete parameter setting. A broad estimate for α_2 is $\alpha_2 = 1$, corresponding to a time constant for Francesco's love of one year and to the following parameter setting:

$$\begin{array}{lll} \alpha_1 = 3 & \alpha_2 = 1 & \varepsilon = 0.1 \\ \gamma_1 = \beta_1 = 1 & \gamma_2 = \beta_2 = 5 & \mu = 100 \\ x_2^* = \delta = 1 & A_1 = 2 & A_2 = -1. \end{array} \tag{13.14}$$

13.4 The L-F cycle and its validation

Equations (13.6-13.8) with the parameter values (13.14) can be numerically integrated for a period of 21 years, starting on the day (6 April 1327) when Laura and Francesco met for the first time and ending on the day she died (6 April 1348). The selected initial conditions are

$$x_1(0) = 0; \qquad x_2(0) = 0; \qquad z_2(0) = 0.$$

The first two are obvious, but the third is also plausible as Petrarch did not write any relevant lyrics before 1327. The results of the numerical integration (shown in Figure 13.3) are qualitatively in full agreement with the "Canzoniere" and with the analysis of Frederic Jones. After a first high

Modeling Love Dynamics

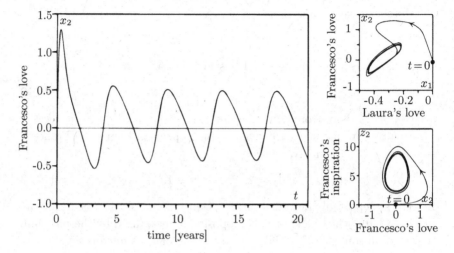

Fig. 13.3 Time evolution of Francesco's love and projections of the trajectory of system (13.6-13.8) with zero initial conditions and parameter values (13.14).

peak, Francesco's love tends toward a regular cycle characterized by alternate positive and negative peaks. Moreover, $x_1(t)$ and $z_2(t)$ tend toward a cyclical pattern. At the beginning, the inspiration of the poet rises much more slowly than his love and then remains positive during the entire period. This might explain why Petrarch writes his first poem more than three years after he has met her, but then continues to produce lyrics without any significant interruption. By contrast, Laura's love is always negative. This is in perfect agreement with the "Canzoniere," where Laura's reactions are repeatedly described as adverse. For example, in sonnet XXI, the poet calls her *"dolce mia guerrera"* [*my sweet enemy*], while in sonnet XLIV he says

né lagrima però discese anchora   *and still no tears your lovely eyes assail,*
da' be' vostr'occhi, ma disdegno et ira.   *nothing as yet, but anger and disdain.*

A comparison of Figure 13.2 with Figure 13.3 shows that the period of the simulated cycle is slightly longer (about 20%) than that identified by Jones. We must therefore increase α_2 by 20% and repeat the simulation for the new parameter setting:

$$\begin{array}{lll} \alpha_1 = 3.6 & \alpha_2 = 1.2 & \varepsilon = 0.12 \\ \gamma_1 = \beta_1 = 1.2 & \gamma_2 = \beta_2 = 6 & \mu = 100 \\ x_2^* = \delta = 1 & A_1 = 2 & A_2 = -1. \end{array}$$ (13.15)

Because of (13.9-13.13) the new solution can be simply obtained by compressing the old one in time by 20%. The new time pattern for Francesco's love is shown in Figure 13.4 together with the grades given by Jones to the 23 dated poems (see Figure 13.2); obviously, the grades are assumed to be proportional to Francesco's love. The fit is very good—actually, as good as can usually be obtained when calibrating models in the physical sciences. Moreover, the fit could be further improved by slightly modifying the parameter values. But we do not show results along this line because we do not want to give the impression that we believe that Petrarch produced his lyrics like a rigid, deterministic machine. Nevertheless, we can conclude that the L-F model with the parameter setting (13.15) strongly supports Frederic Jones's analysis.

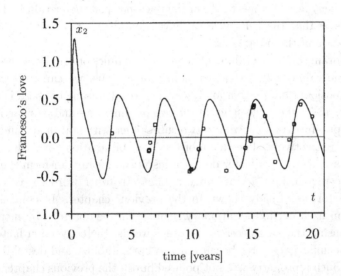

Fig. 13.4 Time evolution of Francesco's love computed with model (13.6-13.8) and parameter values (13.15). The points are estimates of Jones (1995) concerning 23 dated poems (see Figure 13.2).

13.5 Reliability of the results and discussion

Just as in Chapters 9 and 10, we now check that the results we have obtained are not the product of some perverse combination of the values given to the parameters of the model. For this, we determine what happens to the L-F cycle discussed in the previous section when the functions and parameters involved in the model are perturbed. This can be done with a brute force approach, *i.e.*, by repeatedly simulating the model for a great number of randomly generated parameter values and functions. Of course, in so doing one should carefully avoid using perturbations that are too heavy.

A much more precise and effective way of checking the reliability of the cycle is to perform a systematic bifurcation analysis with respect to the parameters. This was done in Rinaldi (1998a) where a one-dimensional bifurcation analysis was carried out numerically with respect to each parameter. These analyses show that the L-F cycle can eventually disappear through a Hopf bifurcation. This means that when the parameters are varied the cycle can shrink and finally be replaced by a stable equilibrium. The variations of the parameters needed to obtain this Hopf bifurcation are relatively large (at least 20% of the reference parameter values (13.15)). This means that the L-F cycle we have identified is not due to a critical combination of the parameters.

To obtain some general insights on the dynamics of couples where there is one individual with extra emotional dimensions, we can also perform two-dimensional bifurcation analyses. The result of one of these is shown in Figure 13.5, where the space of the two parameters ε and μ is partitioned by a Hopf bifurcation curve into two subregions, with stationary and cyclic regimes, respectively. The shape of the Hopf bifurcation curve shows that an increase of ε, *i.e.*, a faster decay of inspiration, can transform the cyclic romantic regime into a stationary regime. In other words, ε is a stabilizing factor, as already shown in the previous chapter. In contrast, the transition from cyclic to stationary regimes can be obtained by increasing or decreasing the parameter μ. In other words, the conversion factor is a rather peculiar parameter because it can both stabilize and destabilize the couple. Such a property was not pointed out in the previous chapter where μ was found to be only stabilizing. As already shown in Chapters 9 and 10, this should not be considered a surprise in the context of dynamical systems. For example, even some linear feedback systems can lose their stability by increasing and also by decreasing the so-called gain of the feedback loop. And, more generally, there are plenty of examples in real life where something can go wrong if something else is done too strongly or too weakly.

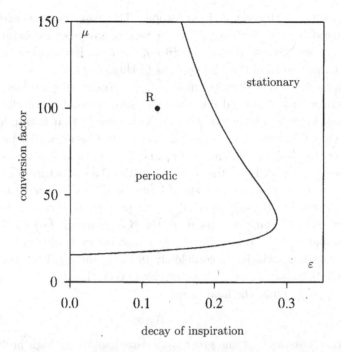

Fig. 13.5   Hopf bifurcation curve in the parameter space (ε, μ). Point R corresponds to the reference parameter values (13.15).

13.6   The great conjecture

Born in Arezzo (Tuscany) in 1304, Francesco Petrarch, perhaps the most celebrated love poet in the Western culture, is undoubtedly an extremely cultivated person. He is a pure scholar interested in history and letters, but he is not attracted by an academic career (he refuses a chair at the University of Florence, offered to him by Boccaccio). He is more attracted by a diplomatic and political career and, indeed, he is appointed *cappellanus continuus commensalis* by Cardinal Giovanni Colonna. This ecclesiastic appointment brings him frequently to Avignon, at the time the residence of the Popes, as well as to many other towns and courts around Europe. During his visits to these centers, Petrarch has plenty of opportunities to influence princes, governors, teachers, and diplomats through meetings, discussions, parties, and other events.

In that same period, Europe evolved culturally after centuries of dormancy, from early medieval symbolic practices to modern humanistic and artistic modes of thought and expression. This coincidence suggests the idea, shared by numerous scholars and here called *great conjecture*, that Petrarch is one of those responsible (if not the main initiator) of the spectacular transition from the Middle Ages to Humanism.

Frederic Jones has already supported this great conjecture because he has shown with his detailed linguistic and stylistic analysis of all the poems of his chronologically reconstructed "Canzoniere," that Petrarch's style evolved smoothly from medieval to humanistic. But his reconstruction relies heavily on the assumption that Laura and Francesco were really involved in an amorous cycle. Thus, the probability that the conjecture is correct is the probability P that the feelings of Laura and Francesco were periodic, which is equal to $(1 - Q)$, where Q is the probability that the cycle did not exist. But (i) some statements in the "Canzioniere," (ii) the analysis of Jones, and (iii) the L-F model, suggest that the cycle did exist. If these three suggestions, which are completely independent one from the other, have probabilities q_i, $i = 1, 2, 3$, to be wrong, then the probability Q is the product $q_1 q_2 q_3$. Thus, the final result is

$$P = 1 - q_1 q_2 q_3.$$

This formula shows that the great conjecture has a very high probability of being correct, even if the probabilities q_i are only slightly smaller than 0.5. If, for example, we conservatively assume that $q_1 = q_2 = q_3 = 0.45$, we obtain $P > 0.9$. In conclusion, the analysis presented in this chapter is relevant not only because it shows that a love story involving one individual with an extra emotional dimension can be successfully described with a mathematical model, but also, and perhaps more importantly, because it strongly reinforces the credibility of the great conjecture.

Chapter 14

Triangular love stories and unpredictability

In this chapter we focus on triangular relationships and show that conflict and jealousy can easily trigger unpredictability. For this, we first identify the structure of all possible triangles with one central individual involved in two romantic relationships. The two non-central individuals can be aware or not of the existence of the triangular relationship, and hence be jealous or not, while the central individual can or cannot feel the conflict between the involvements for the two lovers. Combining these features in all significant ways, we obtain six structurally different triangles that are analyzed separately.

Triangular relationships are often rooted in married couples in which one of the two individuals becomes the center of a triangle on the appearance of a third individual in the game. If the triangle tends toward an equilibrium in which the central individual has unbalanced feelings in favor of the new lover, then the most probable outcome is divorce and formation of a new couple. If, in contrast, the imbalance is in favor of the former partner, then the rupture of the triangle formally closes an extra-marital affair phase.

Although definitely less frequent, triangles in which the two basic couples would have, in isolation, romantic ups and downs, are very intriguing and interesting from a speculative point of view. Indeed, our analysis, based on linear and nonlinear systems theory, shows that these triangles can be characterized by chaotic regimes and hence give rise to unpredictable love stories. Moreover, there are more chances for this to be true if there are more interconnections among all involvements, i.e., if conflict and jealousy are more relevant. This result is consistent with classical rules of thumb that emerged in mathematical modeling after decades of experiences and is quite interesting in the context of love psychology because it gives theoretical support to the interest that psychoanalysts often have in minor details.

No new mathematical prerequisites are needed for reading this chapter.

14.1   Triangular relationships

We have seen in Chapter 11 that chaotic environments entrain couples in turbulent and unpredictable romantic regimes. Actually, we have proved that romantic chaos can emerge even if the environment is simply periodic. In other words, a couple can be involved in an unpredictable regime even if the psychological traits of the two individuals are absolutely normal and the environment varies seasonally. Then we have seen in Chapter 12, without, however, presenting specific examples, that individuals involved in dyadic relationships can also evolve toward unpredictable romantic regimes in constant environments, provided that at least one of them is endowed with a second emotional dimension (like poetic inspiration). Of course, this second dimension can still be a romantic dimension, as in the case of triangular relationships.

Triangular relationships are very frequent (they are experienced, consciously or not, by more than 60% of the American population (Pam and Pearson, 1998, p.149)) and very diversified. They go from extreme *Oedipal triangles* (Johnson, 2010, p.6) to relatively rare *ménages à trois* involving any assortment of sexes (Berne, 1970, p.173). However they more often describe the phase of crisis of a married couple where one of the two individuals becomes the center of a triangle on the appearance of a third individual in the game. The evolution of the feelings in a triangle is a consequence of the individual psychological traits of its components and can therefore be virtually described by a mathematical model of the same kind used until now. Being a dynamical system, the triangle converges in a relatively short time toward a romantic regime that can be very different from case to case. When this regime is an equilibrium, the central individual often has quite unbalanced feelings. If the unbalance is in favor of the new lover, then the most probable outcome is divorce and the formation of a new couple. Hopefully, this new couple is formed with the agreement of all parties, as in the successful cinematographic interpretation of the Arthurian legend starring Julia Ormond as Queen Guinevere, Sean Connery as King Arthur, and Richard Gere as Lancelot. But often, in particular in legends and ancient poems and plays, the successful formation of the new couple is tragically prevented by suicides or murders. This is, for example, the case (with two murders) of Francesca, Gianciotto, and Paolo described by Dante Alighieri and also the case (with one murder and two suicides) of Juliet, Paris, and Romeo described by William Shakespeare. If, in contrast, the imbalance is in favor of the former lover, it is highly probable that the triangle will break and the former monogamic relationship be reestablished. This is frequently the case when one individual has temporary extramarital affairs, like those described by Auguste Flaubert in his masterpiece "Madame Bovary". Triangles characterized by stationary regimes have, in a sense, simple dynamics and that is why we do not analyze them here, even if they are important from a social point of view. We firmly believe, however, that they will attract the attention of scientists in the near future. In contrast, we concentrate on triangles with turbulent (periodic, quasi-periodic, or chaotic) romantic regimes, which are much more intriguing and interesting from a speculative point of view, and we discuss in the next chapter a well documented case study. In perspective, our analysis can be considered as a sound frame for the study of other triangular relationships like those involving the family or groups of friends. Extensions to the more complex case of the so-called *rectangular love stories* (like those described in "A Midsummer Night's Dream" by Shakespeare, or in "Così Fan Tutte" by Mozart) are also possible.

The main property we wish to point out is that triangular love stories with non-stationary regimes are particularly stressful for at least one of the individuals, who, in the end, is the one most likely to break up the relationship. A possible reason for the emergence of this stress is the impossibility of predicting the future of the relationship. In fact, unpredictability is undoubtedly a source of pain: this is why people do not go blindfolded on a roller coaster or why the torture of the Chinese drop (that falls on the head of the victim at random) is so painful. Thus, our thesis is that nonstationary triangular love stories are often unpredictable. To support this thesis we use ODE models of the standard kind, based on oblivion and reactions to love and appeal. But we also add individual characteristics as awareness of the triangular relationship, jealousy, and conflict in the individual involved in two relationships. Of course, these do not include all the psychological traits relevant in a triangle. For example, in the next chapter we study a case in which one of the two lovers is pleased by the existence of another pretender. The ways we use here for modeling jealousy and conflict are also not the only ways possible. Thus, the catalog of triangles that we discuss is far from complete. Nevertheless, it is sufficiently rich to allow us to show, on the basis of known theories, similarities with other investigations, and the *rules of thumb* that have emerged during the history of mathematical modeling, that unpredictability is highly probable in triangles with conflict and jealousy.

14.2   A catalog of triangles

Let us first consider triangles in which only one of the three individuals, say 1, has two lovers. This individual is often called central. The simplest case is when the two lovers of the central individual are unaware of the existence of the triangle or, equivalently, when they are absolutely not influenced by its existence. Under these conditions, individual 1 has two feelings, x_{12} and x_{13}, that measure the involvement that she/he has for the two lovers. In contrast, individuals 2 and 3 have only one feeling, say x_{21} and x_{31}, that measures their interest for the central individual. If individual 1 does not suffer any conflict from having two lovers, then the model has, *a priori*, the following structure

$$\begin{aligned}
\dot{x}_{12} &= f_{12}(x_{12}, x_{21}) \\
\dot{x}_{21} &= f_{21}(x_{21}, x_{12}) \\
\dot{x}_{13} &= f_{13}(x_{13}, x_{31}) \\
\dot{x}_{31} &= f_{31}(x_{31}, x_{13}).
\end{aligned} \qquad (14.1)$$

The first argument of each function is due to the oblivion process (plus, possibly, synergism or platonicity), while the second argument interprets the reaction to the partner's love. The first two ODEs are completely disconnected from the others and describe the evolution of the couple (1-2) in agreement with the theories discussed in Part I of this book. Of course, the same holds true for the last two ODEs that mimic the behavior of the couple (1-3).

The structure of model (14.1) and of all other possible triangles is clearly visualized by the so-called influence graphs in which each node (ij) represents the interest x_{ij} of i for j and the arcs entering each node (ij) represent the direct influence of all other feelings on x_{ij}. For example, the structure of model (14.1) is depicted in Figure 14.1a which shows, unambiguously, that the triangle is nothing but two disconnected couples. In a sense, we have already studied a triangle of this kind, because in Chapters 9 and 10 we have discussed the love stories of Kathe and Jules and of Kathe and Jim as if they were fully independent. But this was only a simplifying assumption and the next chapter is devoted to the study of a more realistic model of the triangle Kathe-Jules-Jim.

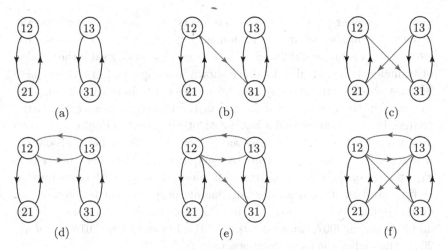

Fig. 14.1   Influence graph of triangles in which only individual 1 (called central) has two lovers (individuals 2 and 3). Node (ij) represents the interest x_{ij} of individual i for individual j. The presence of an arc from node (hk) to another node (ij) means that the feeling x_{ij} is directly influenced by the feeling x_{hk}. Black arcs interpret reactions to love, blue arcs conflict in individual 1 and red arcs jealousy. The characteristics of the triangles are: (a) no conflict, no jealousy; (b) no conflict, 3 is jealous; (c) no conflict, 2 and 3 are jealous; (d) conflict, no jealousy; (e) conflict, 3 is jealous; (f) conflict, 2 and 3 are jealous.

   If one of the two lovers, say 3, is aware of the existence of the triangle and is jealous, the feeling x_{31} is also influenced by the interest x_{12} that the central individual has for lover 2. This means that in model (14.1) the last ODE should be modified and the new function f_{31} should have x_{12} as extra argument. This is clearly pointed out by the influence graph of Figure 14.1b. Of course, if both lovers are aware and jealous, then the influence graph is that of Figure 14.1c.

   It can happen (actually it is often the case) that the central individual is not capable of isolating her/his two romantic spheres, in which case, there is an interaction between x_{12} and x_{13} described with two suitable functions f_{12} and f_{13} in model (14.1). This interaction can be due to the difficulty of withstanding two high and independent involvements. For example, if individuals are heterosexual and the central one is attracted by monogamy, then it is natural to expect that the oblivion processes of 1 are not the same for the two lovers and vary in time, penalizing the lover for which the involvement is lower. In the next chapter, Kathe is assumed to follow this adaptive behavior. In conclusion, if the central individual is in conflict, the triangle has the structure shown in Figure 14.1d, 14.1e, or 14.1f, depending upon the number (0, 1, or 2) of jealous lovers. The most complex triangle, in terms of number of arcs in its influence graph, is that of Figure 14.1f where 1 is in conflict and 2 and 3 are jealous. As already said, the catalog of triangles with central individuals shown in Figure 14.1 is not complete because conflict and jealousy can be modeled in alternative ways. For example, in the first model discussed in the literature (Sprott, 2004) the triangle has a structure similar but not identical to that in Figure 14.1c (two arcs are not oriented in the same way). Moreover, the model is absolutely non-generic because the structure of its equations implies that the feelings x_{12} and x_{13} of the central individual become equal but with opposite sign as time goes on: Indeed a rather degenerate property. For this reason, we do not report here results concerning this model or its extensions (Ahmad and El-Khazali, 2007; Koca and Ozalp, 2014; Liu and Chen, 2015; Xu *et al.*, 2011) that suffer the same degeneracy.

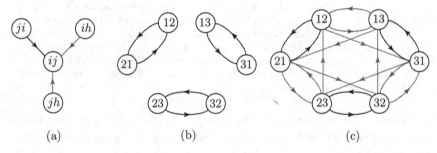

(a)   (b)   (c)

Fig. 14.2   Influence graph of triangles in which each individual has two lovers (*Ménages à trois*). Node (ij) represents the interest x_{ij} of individual i for individual j. The presence of an arc from node (hk) to another node (ij) means that the feeling x_{ij} is directly influenced by the feeling x_{hk}. Black, blue, and red arcs interpret reaction to love, conflict, and jealousy, respectively. (a) The three arcs entering in each node; (b) the influence graph in the extreme case of no conflict and no jealousy; (c) the influence graph when each individual is in conflict and is jealous.

A more complex class of triangles, identified with the French expression *ménages à trois*, is that in which each individual is in romantic/sexual relationship with the others. In this case, each individual i has a feeling x_{ij} for individual j and a feeling x_{ih} for individual h, for a total of six different feelings. This means that the influence graph has six nodes. Using the same colors—black, blue, and red—used in Figure 14.1 to identify reaction to love, conflict, and jealousy, the arcs entering in a generic node (ij) of the influence graph are three, as shown in Figure 14.2a. Imposing this pattern of arcs in each node (assuming that each individual is in conflict and jealous) one obtains the influence graph of Figure 14.2c. Of course, it is absolutely not said that conflict and jealousy are affecting all individuals, *i.e.*, that all blue and red arcs are present in the graph. The simplest case, in which conflict and jealousy are totally absent, is described by the graph of Figure 14.2b, which shows that this triangle can be studied by analyzing three independent dyadic love stories.

14.3 Jealousy, conflict, and unpredictability

Studying the unpredictability of the future behavior of all types of triangles for all possible combinations of the characters of the three individuals (secure *vs.* insecure, unbiased *vs.* biased) would require an enormous effort and an entire book for the description of the results. Here we limit our attention to some significant triangles (precisely those shown in Figure 14.1) and try to derive general (and hence weak) conclusions without specifying the characters of the individuals involved. This means that our target is very ambitious: discuss the unpredictability of triangles on the sole basis of their structure (*i.e.*, the topology of the influence graph). We can anticipate the result of our analysis by saying that the chances of unpredictability increase from (a) to (f).

Let us start by recalling a few basic things concerning unpredictability. Mathematically speaking, unpredictability is a property present in systems with chaotic regimes. Thus, unpredictability is revealed by the strict positivity of the Largest Lyapunov Exponent (LLE), as briefly discussed in the Appendix and in Chapter 11. It is worth noting that the LLE can be easily computed if a model is available. In principle, it can also be computed from long and reliable time series, which, however, are not available in the context of love dynamics. Alternatively, but equivalently, the sign of the LLE can be established (through simulation) by looking at the geometry of the attractor on a suitable Poincaré section. A dynamical system modeled by n ODEs has n Lyapunov exponents which, physically speaking, describe the sensitivities of the state of the system to perturbations of the initial conditions. If the largest of them, *i.e.*, the LLE, is positive, then two infinitesimally close initial conditions give rise to evolutions that diverge one from the other at an exponential rate (in the so-called stretching phase), before reconverging (in the so-called folding phase), thus creating the conditions for a new stretching, and so on. The stretching mechanism can easily be discovered through simulation, as shown in Figure 14.3, where the

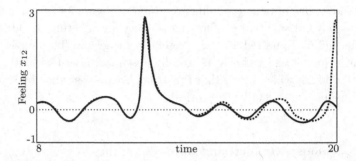

Fig. 14.3   The stretching phase of a chaotic system: evolution of the feeling x_{12} of the central individual for lover 2 in the triangle studied by Sprott (2004). The two initial conditions look the same because their distance is smaller than the dimension of the pixel. The time series are extracted from Figure 5 in Sprott (2004).

first simulation carried out for a triangle (Sprott, 2004) is reported. But the evolution of a triangle can be, in practice, quite complex even if the romantic regime is quasi-periodic (LLE=0), because in this case, as explained below, the central individual can be entrained into aperiodic behavior.

Let us start our analysis with the triangles from which conflict and jealousy are absent, *i.e.*, with Figure 14.1a. In this case the triangle is composed of two independent couples. If we assume that the couples are as in Part I of this book, then each couple can tend either toward an equilibrium or toward a limit cycle. As already mentioned, the less frequent but by far most intriguing case is that of two couples with cyclic regimes of period T' and T'', respectively. If the ratio of the two periods is rational, *i.e.*, if

$$\frac{T''}{T'} = \frac{n_1}{n_2} \qquad (14.2)$$

with n_1 and n_2 integer and coprime, then in an interval of time length

$$T = n_1 T' (= n_2 T'')$$

the first couple goes through n_1 complete cycles, while the second couple goes through n_2 complete cycles. This means that the triangle behaves also periodically but with a period T which is the least common multiple of the two periods T' and T''. Thus, the period T can be very large, in particular if T' and T'' are not very different, because in that case n_1 and n_2 are large.

In principle, equation (14.2) should never (or almost never) be satisfied because the ratio of two independent real numbers (T' and T'') has zero probability of being a rational number. In the generic case (that is, in the

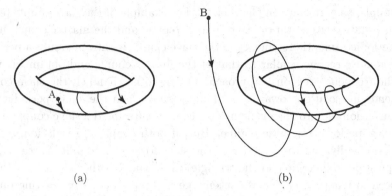

Fig. 14.4   Quasi-periodic regime: visualization of a torus in a three-dimensional space. (a) A trajectory (starting from point A) lying on torus; (b) A trajectory tending toward (starting from point B) a torus.

case of probability 1) the ratio between T' and T'' is an irrational number. In such a case, the triangle asymptotically tends toward a quasi-periodic regime, *i.e.*, in the space of the feelings, the trajectories tend toward a torus. A visualization of the torus is not possible because the space of the feelings is four-dimensional. However, one can imagine that this torus is a generalization of the standard torus of classical geometry shown in Figure 14.4. A trajectory starting from a point on the torus (point A in Figure 14.4a) remains forever on the torus without, however, returning to the starting point. Indeed, if this did occur, property (14.2) would hold and the behavior would be periodic (cycle on torus). This means that the trajectory densely fills the torus, as time goes on. In contrast, a trajectory starting from a point which is not on the torus (point B in Figure 14.4b) tends asymptotically toward the torus. Thus, we can conclude that in the absence of conflict and jealousy (Figure 14.1a) we cannot have unpredictability because the asymptotic regime is not chaotic (or, equivalently, because the LLE is not positive). But, in any case, we have a quite complex romantic regime which is generically aperiodic.

In the case of Figure 14.1b, *i.e.*, when one of the two lovers is jealous, the two couples are not independent because couple (1-2), the master couple, influences couple (1-3), the slave couple. This means that the love story of the master couple is not influenced by the presence of the slave couple, while the slave couple suffers the influence of the master couple. If jealousy is weak, the evolution of the love story of the slave couple can be predicted from that of the master couple by studying the linearized slave couple, as performed in Chapter 11. For example, if the slave couple tends, in isolation, to a stationary romantic regime and the master couple has a periodic romantic regime, then the slave couple is entrained into a periodic romantic regime similar to that of the master couple and obtainable with the frequency response approach. On the other hand, if the slave couple tends in isolation toward a periodic regime, then the small periodic perturbation, due to the cyclic master couple, entrains the slave couple into a quasi-periodic romantic regime. But, if both couples, in isolation, behave periodically and the jealousy of the slave lover is gradually increased, the quasi-periodic regime of the triangle can break. In other words, as already seen in Chapter 11, chaos can emerge if the periodic stress acting on the slave lover is sufficiently high. When this occurs, the behavior of the triangle becomes unpredictable and the individual who suffers this unpredictability most is the jealous slave. From what we said in Chapter 11, we can also expect that the level of jealousy required to trigger strict unpredictability is lower if, in isolation, the two couples oscillate at similar frequencies. In other words, if slave and master romantic clocks beat at comparable frequencies, then it is highly probable that even a not-too-jealous slave will be incapable of predicting the evolution of the love story. In such a case the jealous slave will be likely to interrupt the triangular relationship.

In the case of Figure 14.1c, namely, when everybody is aware of the existence of the others and both lovers are jealous, the two couples are mutually interconnected because the master-slave connection is substituted by a feedback connection. The behavior of systems of this kind is particularly complex, even in the linear case, because feedbacks are the primary cause of instabilities. Thus, chaos has higher chances of existing than in the case of a single jealous lover. An interesting feature that can emerge in these triangles for suitable forms of jealousy is *intermittency*, namely, the possibility that the chaotic regime is a concatenation of long and random periods of time during which each couple alternately plays the role of slave and of master.

In the case of Figure 14.1d, namely, when there is no jealousy and individual 1 is in conflict, the two couples are mutually interconnected so that the romantic regime should again be expected to be quasi-periodic if the conflict is weak. The two couples can also lock their behaviors (in such a case, in the space of the feelings, the attractor is a cycle on the torus) but this is a rather exceptional phenomenon that occurs only in very tiny regions of the space of the traits, called Arnold's tongues (see Appendix). Increasing the conflict, a sequence of complex bifurcations, characterizing the so-called torus destruction route to chaos, can transform the torus into a genuine strange attractor. However, this transition is not guaranteed and requires, in any case, a strong interconnection between the two couples, *i.e.*, a high conflict in the central individual. For example, in the triangle discussed in the next chapter the regime remains quasi-periodic even for high values of the conflict.

Finally, in the cases of Figures 14.1d and 14.1e, we would expect romantic regimes to fall easily into chaos because jealousy and conflict are copresent and each of them, alone, has the power of generating chaos. This is also confirmed by the case study presented in the next chapter.

The analysis of the triangles described in Figure 14.1, namely, those in which only one individual has two lovers, can be summarized by saying that if the two couples were to have, in isolation, romantic ups and downs, then the evolution of the feelings in the triangle would be generically unpredictable. As unpredictability is, undoubtedly, a serious source of stress, the consequence is that, in general, triangles of this kind should be expected to break (even dramatically) after some time. On the other hand, the triangular love stories tending toward a stationary regime are also expected to be interrupted by the central individual if the feelings for the two lovers are unbalanced. Thus, the triangles that seem to have high chances of persisting for a long time are those characterized by stationary and balanced regimes, *i.e.*, those in which, after an initial transient, the interests of the central individual for the two lovers are comparable because the two lovers are suitors of radically contrasting personalities: one of a girl next door or nice guy type, and the other physically attractive but a potentially hazardous type of person.

We could now make similar considerations for the triangles in which each individual has two lovers (ménages à trois). For this, we should consider a great number of influence graphs, starting with Figure 14.2b (no conflict and no jealousy) and ending with Figure 14.2c (conflict and jealousy in each individual). We do not begin this discussion because it is actually superseded by the next section. The only remark we wish to make concerns Figure 14.2b, namely, the case in which there are three independent love stories. If these love stories are periodic and the ratios of their periods T', T'', and T''' are rational numbers, then the triangle has periodic behavior, with period T given by the least common multiple of the three periods. Thus, if the three periods are comparable, T can be very large and this implies that the romantic regime is periodic with a very long period. This particular property suggests that "ménages à trois" are generically more complex than the triangles shown in Figure 14.1.

14.4   Conclusions and discussion

We now present a few general comments that could be useful for extending the analysis we have carried out for triangles to more general structures of interpersonal relationships.

In this chapter we have tried to identify the chances of emergence of chaos and hence unpredictability in networks of social interactions, limiting our analysis to triangles. More technically, the task has been to derive information on the most probable asymptotic behavior of triangular relationships by looking only at their structure, without specifying important features of the individuals involved. The characteristics that define the structure of the triangle are the existence of conflict in the central individual and of jealousy in the two lovers. The chances of emergence of other properties like the existence of alternative romantic regimes, the possibility of catastrophes (*i.e.*, switches from one romantic regime to another) and the synchronization of the feelings, could also be discussed in terms of the sole structure of the triangle, but this is not done here because there are not yet significant theoretical results on these issues. Rigorously speaking, the ambitious idea of connecting the properties of a triangle (or, more in general, of a network) to its structure is justified only in particular classes of systems because, in general, the individual properties and the values of the parameters also matter. Only in the case of positive linear systems (Farina and Rinaldi, 2000), also called cooperative systems because each variable has a positive influence on all others, does the structure of the influence graph determine all basic properties. But unfortunately, our triangles are neither linear nor cooperative because the reactions to love are nonlinear and can be decreasing (and even negative). However, on the basis of many modeling experiences, carried out in various fields of science, a few rules of thumb have emerged that can be used to tentatively answer our questions.

The first studies on the complexity of dynamical systems (Gardner and Ashby, 1970; May, 1972; Siljack, 1974), performed on positive systems, have come to the general conclusion that the probability of complex behaviors increases with the number of interactions. Different experiences, in particular those performed in the field of industrial automation, suggest a different rule, namely, that complexity increases with the number of elementary cycles present in the influence graph. It is interesting to note that our conclusions are consistent with both rules. In fact, our analysis has shown that the possibility of having chaos (*i.e.*, unpredictability) in the triangles of Figure 14.1 increases (or, better, does not decrease) from (a) to (f), while the number of arcs and of elementary cycles present in the graphs are

	(a)	(b)	(c)	(d)	(e)	(f)
# arcs	4	5	6	6	7	8
# cycles	2	2	3	3	4	6

Another interesting discovery emerging from our analysis is that chaos is the result of interferences between oscillatory mechanisms of comparable strength and frequency. This conclusion is also in line with other investigations carried out in completely different sectors. For example, in ecology each consumer population has a favorite resource but can also feed on a secondary species, which, in turn, can be the favorite resource for another consumer. Thus, complex food webs are naturally described as consumer-resource pairs interconnected through feeding preferences. A model of two consumers competing for two resources has therefore the same structure as the models considered in this chapter. This is of great potential interest because some of the general results obtained in mathematical ecology (Hastings and Gross, 2012), in biochemistry (Goldbeter, 1996; Novák and Tyson, 2008; Zhang *et al.*, 2012), and in the study of coupled oscillators (Pikovsky *et al.*, 2001) could guide the modeling of complex interpersonal relationships.

A last remark is that our analysis confirms a general belief in the study of dynamical systems, namely, that small details (here jealousy and conflict) can promote complex behaviors. In the specific context of interpersonal relationships, this justifies the interest that psychoanalysts have in apparently minor aspects of the characters.

Chapter 15

Kathe, Jules, and Jim

In this chapter we study the triangular love story involving Helen Grund, a brilliant and charming journalist, her husband Franz Hessel, a profound but shy German writer, and his best friend Henry-Pierre Roché. The story is described in detail in the autobiographic novel "Jules et Jim", written in 1953 by Roché, and popularized in 1961 by the homonymous film of François Truffaut—one of the prominent directors of the "*Nouvelle Vague*". In the novel, and in the film, Helen, Franz, and Henry-Pierre are Kathe, Jules, and Jim, respectively (see Figure 15.1). The story has two distinctive features. First, the two friends are introduced to Kathe practically at the

Fig. 15.1   Flier of the film "Jules et Jim" showing, from left to right, Jim, Jules, and Kathe.

same time, so that all initial feelings can be set to zero. This simplifies the analysis and eliminates the ambiguities concerning possible alternative attractors. Second, although she is permanently involved with both friends, Kathe is always in a monogamic relationship and changes partner each time, there is an inversion in her preference ranking.

Since the basic psychological traits of the three individuals have already been discussed in Chapters 9 and 10, here we concentrate on the characteristics identifying the structure of the triangle. In Roché's novel we find clear evidence of the complaisance of Jules, who is pleased when Kathe is more in love with Jim, and of the jealousy of Jim. To interpret the synergic character of Kathe, we also assume that she is in conflict and forgets less quickly the lover she is more involved with. Under these assumptions, the resulting structure of the triangle is one for which, in the previous chapter, we have predicted chaotic behavior if the interactions between the two couples are sufficiently strong. This is, indeed, what we find by simulating the model for all reasonable combinations of the parameters measuring complaisance, jealousy, and conflict. Moreover, a great number of details in Roché's novel allow us to validate the model, not only with respect to qualitative features of the love story, but also with respect to quantitative information like the dates of the seven partner changes in the twenty years of concern.

Finally, the results of the analysis are used to highlight the genius of François Truffaut who, two years before the appearance of the first scientific paper on chaos, systematically used the metaphor of "stretching and folding" to explain the turbulence of the love story.

No new mathematical prerequisites are needed for reading this chapter. Extra details can be found in Dercole and Rinaldi (2014).

15.1 The story

As already mentioned in Chapters 9 and 10, "Jules et Jim" is the first novel by Henri-Pierre Roché, published in 1953 (Roché, 1953) when he was already 74. It is an autobiographical novel describing the triangular relationship involving Helen Grund (Kathe), her husband Franz Hessel (Jules), and Henri-Pierre Roché himself (Jim), an intimate friend of Franz Hessel. The story begins in Paris a few years before the First World War, when the two friends are introduced to Kathe, a German girl who has come from Berlin to practice in the Parisian painters' ateliers. Except for the war period, that sees the two friends on opposite sides, Kathe, Jules, and Jim mostly live together in Paris, Berlin, or in the German countryside, for a time-span of about 20 years.

There are two slightly different descriptions of the love story. One is Roché's novel, and the other is the homonymous 1961 film of François Truffaut. Actually, some interesting extra information can also be found in the diaries and letters of Henri-Pierre Roché, Franz Hessel and Helen Grund, which, however, are not used in the following. For our modeling purposes, the two descriptions share the same kind of love dynamics. In the next sections (as in Chapters 9 and 10), we base our modeling assumptions on Roché's novel and use selected scenes of the film only at the end of the chapter to highlight the contribution of Truffaut.

15.2   The free-love hypothesis

Roché's novel is interesting for two reasons. First, because it is autobiographical and thus a reliable source of information, which allows us to validate a mathematical model against a real love story. Second, because it conveys the central idea of Roché's philosophy, which can be summarized by saying that we should not try to possess or constrain the people we love, but leave them free to engage in other relationships. This anti-bourgeois ideology, first supported, among others, by Bertrand Russell and known as "free love", was particularly appreciated during the social movements of the sixties. This is possibly why both the novel and the film have been so successful.

As already explained in Chapters 9 and 10, Roché's ideology of free love is important for our purposes, because it suggests that Kathe should be split into two independent persons and that the virtual couples Kathe-Jules and Kathe-Jim should be described separately. Kathe is in fact the character used by Roché to support his philosophy. She basically lives two parallel love stories with no particular internal conflict, as explicitly stated by Roché in the following quote:

> *In her mind, each lover was a separate world, and what happened in*
> *one world was no concern of the others* (p.108)

The two pairwise virtual love stories develop almost independently and the deep friendship between Jules and Jim attenuates possible jealousies:

> *In twenty years Jim and he had never quarrelled. Such disagreements*
> *as they did have they noted indulgently* (p.237)

As a consequence, a first model of the triangular love story is the union of models (9.1, 9.2) and (10.1, 10.2) developed and studied in Chapters 9 and 10, respectively. Model (9.1, 9.2, 10.1, 10.2) is called the "free-love model", because it follows from the "free-love hypothesis" under the extreme assumption of no interference between the two couples. The asymptotic regime of the model is quasi-periodic because the periods of the Kathe-Jules and Kathe-Jim cycles are not commensurable (they are known numerically with finite approximation, but their ratio is most likely irrational; see the discussion in Section 14.3). The triangular love story is described in Figure 15.2 where the first 20 years of transient toward the quasi-periodic regime are shown. Note that all feelings are set to zero at time $t = 0$, since Jules and Jim are together when they first see Kathe.

There are five qualitative features of the love story identifiable in the novel that are well reproduced by the free-love model.

(i) During the first years Kathe is more attracted by Jules (she marries him). In accordance with the notation used in the previous chapter, this corresponds to saying that $x_{12} > x_{13}$ in Figure 15.2;

(ii) At the very beginning of the story, Kathe is more attracted by Jim, who misses a strategic date:

> *If Kate and Jim had met at the café, things might*
> *have turned out very differently* (p. 80)

This is not visible at the scale of Figure 15.2. However, Jules' appeal is lower than that of Jim ($A_2 < A_3$, see Tables 9.1 and 10.1), and this implies that during the very first days of the story the feeling of Kathe for Jules is lower than for Jim ($x_{12}(1) = \gamma_1 A_2$ and $x_{13}(1) = \gamma_1 A_3$ according to the discretized models (9.3, 9.4) and (10.3, 10.4)). But after a couple of weeks Kathe's preference is for Jules and she marries him soon afterwards.

(iii) Jim's ups and downs are more relevant than those of Jules:

> *Jim was easy for her to take, but hard to keep. Jim's love drops*
> *to zero when Kate's does, and shoots up to a hundred with*
> *hers. I never reached their zero or their hundred* (p.231)

This is very visible in Figure 15.2, where x_{31}-oscillations are wider than x_{21}-oscillations.

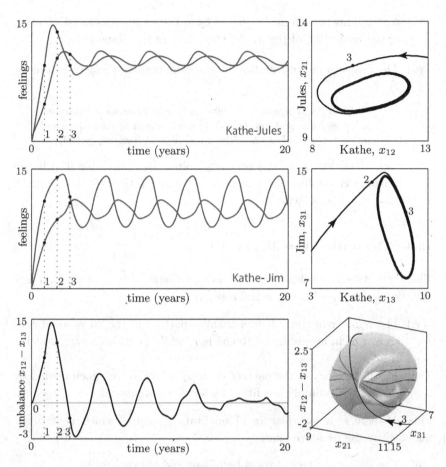

Fig. 15.2 The love stories predicted by the free-love model.

(iv) The drops in interest of Kathe for Jules anticipate those of Jules for Kathe:

> *The danger was that Kate would leave. She had done it once already... and it had looked as if she didn't mean to return... She was full of stress again, Jules could feel that she was working up for something* (p.89)

This is confirmed in Figure 15.2 and follows from Kathe's insecureness with Jules due to Jules' platonicity. When Jules' love for Kathe increases, Kathe's reaction to love becomes negative and causes the inversion in her feeling for Jules (a local maximum of x_{12}). And only after a sufficient drop in Kathe's feelings does Jules also invert his trend. This is even more evident from the counterclockwise rotation of the projection of the model trajectory in the plane (x_{12}, x_{21}).

(v) The drops in interest of Jim for Kathe anticipate those of Kathe for Jim:

> *He himself was incapable of living for months at a time in close contact with Kate, it always brought him into a state of exhaustion and involuntary recoil which was the cause of their disasters* (p.189)

This is analogous to the previous point. But this time Jim is the insecure character, and the rotation of the trajectory in the plane (x_{13}, x_{31}) is clockwise.

15.3   The Kathe-Jules-Jim model

There are two other features that are of quantitative nature and not matched by the free-love model. These are:

(vi) The number of times Kathe changes partner in the 20 years of concern, which, according to Roché is 7, while in the free-love model it is 9.

(vii) The chronology of the partner changes, which is well documented by Roché, is absolutely not fitted by the free-love model.

Finally, there is a last feature of qualitative nature, which is also not matched by the free-love model:

(viii) The triangular love story is turbulent and unpredictable.

A remarkable turbulence in the feelings of the three characters is easily perceptible when reading the novel, and one indeed has the impression that Kathe is quite unstable and difficult to predict, especially in her partner changes. Actually, the uncertainty over the future creates in the triangle (and in the reader) an increasing tension that ceases only when Kathe and Jim commit suicide:

> *Jules would never have again the fear that had been with him since the day he met Kate, first that she would deceive him—and then, quite simply, that she would die, for she had now done that too* (p.236)

Indeed, the dramatic end imagined by Roché can be interpreted as a poetic way of interrupting the unsustainable torture due to the recurrent but unpredictable changes of partner. The available data on the partner changes also confirm that they are irregular.

To match features (vi)–(viii), we now relax the free-love hypothesis, and propose a model for the triangle, called the K-J-J model, in which weak interferences between the virtual couples Kathe-Jules and Kathe-Jim are allowed. On the basis of the discussion in the previous chapter, we should *a priori* expect this perturbation of the free-love model to have high chances of triggering some form of unpredictability.

We first take into account the internal conflict in Kathe and assume that it affects her forgetting process. Specifically, we assume that at any given time she forgets less quickly the lover she is more involved with. This is realized by multiplying, in the free-love model, her forgetting coefficient α_1 (see equations (9.2) and (10.2)) by a factor which is smaller than 1 in the equation for the partner she loves more and greater than 1 in the other, thus obtaining equations (15.1) and (15.2). In order to deviate only slightly from the free-love principles, the conflict parameter ε must be small and positive.

Moreover, Roché explicitly describes specific behaviors on the part of the two friends that violate the rigid principles of free love. Jules is complaisant about Jim—he is pleased when Kathe is with Jim because he realizes this makes her happier. This characteristic, peculiar to Jules, is consistent with his platonic nature (see Chapter 9) and is well described by Roché:

> ... *I'm terrified of losing her, I can't bear to let her go out of my life. Jim—love her, marry her, and let me go on seeing her. What I mean is, if you love her, stop thinking that I'm always in your way* (p.27)

Although jealousy is at odds with the ideology of free love, Jim is slightly jealous of Jules:

> *She bestowed her graciousness on each in turn... and Jim was jealous* (p.97)

To take Jules' complaisance into account, his reaction to Kathe's love is amplified by a factor greater than 1 when she is more in love with Jim, namely, when x_{13} is greater than x_{12} (see (15.3)). Symmetrically, to take Jim's jealousy into account, his reaction to Kathe's love is attenuated by a factor smaller than 1 when she is more in love with Jules, namely, when x_{12} is greater than x_{13} (see (15.4)). For simplicity's sake, Jules' complaisance and Jim's jealousy are quantified by the same positive parameter δ, that must also be small if we wish to avoid large deviations from free-love principles.

In conclusion, the K-J-J model is composed of the following four equations:

$$\dot{x}_{12} = -\alpha_1 \exp(\varepsilon(x_{13}-x_{12}))x_{12} + R_{12}^L(x_{21}) + (1+S(x_{12}))\gamma_1 A_2, \qquad (15.1)$$

$$\dot{x}_{13} = -\alpha_1 \exp(\varepsilon(x_{12}-x_{13}))x_{13} + \beta_{13}x_{31} + (1+S(x_{13}))\gamma_1 A_3, \qquad (15.2)$$

$$\dot{x}_{21} = -\alpha_2 x_{21} + \beta_2 x_{12} \exp(\delta(x_{13}-x_{12})) + (1-P(x_{21}))\gamma_2 A_1, \qquad (15.3)$$

$$\dot{x}_{31} = -\alpha_3 x_{31} + R_{31}^L(x_{13}) \exp(\delta(x_{13}-x_{12})) + \gamma_3 A_1, \qquad (15.4)$$

and differs from the free-love model in the presence of the two small coupling parameters ε and δ. The analytical expressions of the reaction functions and the reference values of the parameters of the K-J-J model are reported in Table 15.1, obtained by merging Tables 9.1 and 10.1.

15.4   Analysis and results

We now validate the K-J-J model (15.1–15.4) against the features discussed above. We keep all parameters (except ε and δ) at the reference values of Table 15.1 and we first look for pairs (ε, δ) for which features (vi) and (vii) are reproduced by the model. For this we fix a dense grid in the (ε, δ) plane and we systematically simulate the model for each point of the grid, always starting from the state of indifference $(x_{12}, x_{13}, x_{21}, x_{31}) = (0,0,0,0)$ and stopping the simulation after 20 years. As already said, we assume that Kathe changes partner, switching from Jules to Jim (or vice versa), as soon as her preference for Jules $x_{12} - x_{13}$ goes from positive to negative (or from negative to positive). The result is that for pairs (ε, δ) in the shaded region in Figure 15.3, Kathe's imbalance $x_{12} - x_{13}$ changes sign seven times in the 20 years of concern. For the particular values of ε and δ corresponding to the white dot in the figure, the predicted love story is depicted in Figure 15.4, which shows that the chronology of the partner changes predicted by the model is in very good agreement with the one documented by Roché (the correlation between the seven instants suggested by the model and those indicated by Roché is 0.97! See the bottom-right panel).

Features (vi) and (vii) are therefore well matched by model (15.1–15.4) for all pairs (ε, δ) in the shaded region in Figure 15.3. For all such pairs, we have then checked that features (i)–(v) are also well reproduced by the model. This is not surprising as ε and δ are small parameters coupling the

Nonlinear functions (specified for non-negative feelings)

Kathe	$R^L_{12}(x_{21}) = \beta_{12} \dfrac{x_{21}}{1 + x_{21}/\sigma^L_{12}} \cdot \begin{cases} \dfrac{1 - ((x_{21} - \tau^I_{12})/\sigma^I_{12})^2}{1 + ((x_{21} - \tau^I_{12})/\sigma^I_{12})^2} & \text{if } x_{21} \geq \tau^I_{12} \\ 1 & \text{if } x_{21} < \tau^I_{12} \end{cases}$ reaction to love $S(x_{1j}) = \begin{cases} s\dfrac{((x_{1j} - \tau^S)/\sigma^S)^2}{1 + ((x_{1j} - \tau^S)/\sigma^S)^2} & \text{if } x_{1j} \geq \tau^S \\ 0 & \text{if } x_{1j} < \tau^S \end{cases} \quad j = 2,3$ synergism
Jules	$P(x_{21}) = \begin{cases} p\dfrac{((x_{21} - \tau^P)/\sigma^P)^2}{1 + ((x_{21} - \tau^P)/\sigma^P)^2} & \text{if } x_{21} \geq \tau^P \\ 0 & \text{if } x_{21} < \tau^P \end{cases}$ platonicity
Jim	$R^L_{31}(x_{13}) = \beta_3 \dfrac{x_{13}}{1 + x_{13}/\sigma^L_{31}} \cdot \begin{cases} \dfrac{1 - ((x_{13} - \tau^I_{31})/\sigma^I_{31})^2}{1 + ((x_{13} - \tau^I_{31})/\sigma^I_{31})^2} & \text{if } x_{13} \geq \tau^I_{31} \\ 1 & \text{if } x_{13} < \tau^I_{31} \end{cases}$ reaction to love

Parameters

Kathe	forgetting	α_1	= 2 [yrs^{-1}]	forgetting coefficient
	reaction to Jules' love	β_{12}	= 8 [yrs^{-1}]	reaction coefficient to love
		σ^L_{12}	= 10	sensitivity of reaction to love
		τ^I_{12}	= 2.5	insecureness threshold
		σ^I_{12}	= 10.5	sensitivity of insecureness
	reaction to Jim's love	β_{13}	= 1 [yrs^{-1}]	reaction coefficient to love
	reaction to appeal	γ_1	= 1 [yrs^{-1}]	reaction coefficient to appeal
	synergism	s	= 2	maximum synergism
		τ^S	= 9	synergism threshold
		σ^S	= 1	sensitivity of synergism
	appeal	A_1	= 20	appeal
Jules	forgetting	α_2	= 1 [yrs^{-1}]	forgetting coefficient
	reaction to love	β_2	= 1 [yrs^{-1}]	reaction coefficient to love
	reaction to appeal	γ_2	= 0.5 [yrs^{-1}]	reaction coefficient to appeal
	platonicity	p	= 1	maximum platonicity
		τ^P	= 0	platonicity threshold
		σ^P	= 1	sensitivity of platonicity
	appeal	A_2	= 4	appeal
Jim	forgetting	α_3	= 2 [yrs^{-1}]	forgetting coefficient
	reaction to love	β_3	= 2 [yrs^{-1}]	reaction coefficient to love
		σ^L_{31}	= 10	sensitivity of reaction to love
		τ^I_{31}	= 9	insecureness threshold
		σ^I_{31}	= 1	sensitivity of insecureness
	reaction to appeal	γ_3	= 1 [yrs^{-1}]	reaction coefficient to appeal
	appeal	A_3	= 5	appeal

Table 15.1 Nonlinear functions and reference parameter values of the K-J-J model.

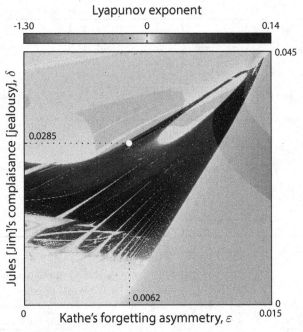

Fig. 15.3   Behavior of the K-J-J model with respect to the coupling parameters ε and δ. The largest Lyapunov exponent is positive (red) for chaotic attractors; zero (yellow) for quasi-periodic attractors and bifurcating cycles; negative (green) for stable cycles. For (ε, δ) in the shaded region the model predicts that Kathe will change partner seven times in 20 years.

two oscillating models (9.1, 9.2) and (10.1, 10.2) composing the free-love model. Thus, all qualitative features matched by the free-love model are also reproduced after the introduction of a weak coupling.

Last, to support feature (viii), we need to show that within a period of 20 years, trajectories originating from similar initial states have the time first to get close to the attractor and second to significantly separate (stretching phase). Indeed, only under these conditions can the three individuals feel the unpredictability of their love story before its end. As for the time needed to reach the attractor, we see from Figure 15.4 that 3 years are largely sufficient. On the other hand, the divergence time of nearby trajectories is of the order of L^{-1}, where L is the largest Lyapunov exponent associated with the attractor. But, for ε and δ corresponding to the white dot in Figure 15.3, the chaotic attractor has $L = 0.07$ yrs^{-1}, so that the characteristic time of divergence of nearby trajectories is about 15 yrs. Thus, in conclusion,

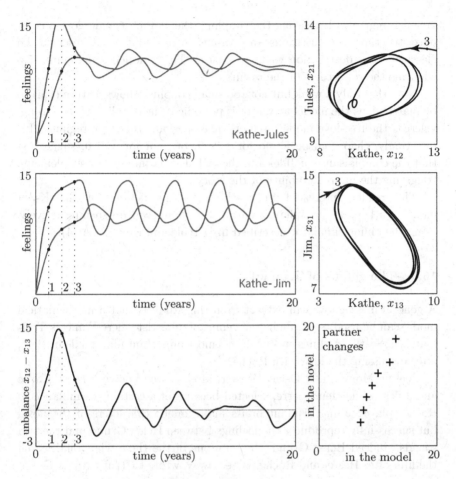

Fig. 15.4   The love stories predicted by the K-J-J model (15.1–15.4) for ε and δ corresponding to the white dot in Figure 15.3.

unpredictability has definitively characterized the love story of Kathe, Jules, and Jim.

The largest Lyapunov exponent L was computed for all pairs (ε, δ) considered in Figure 15.3 (see the color-code), and the result is the typical bifurcation diagram expected for weakly coupled oscillators. For extremely weak coupling, the model attractor is a torus (quasi-periodic regime; see the yellow region close to $\varepsilon = \delta = 0$). Then, for larger coupling, the two oscillators can synchronize (frequency locking) on a cycle on torus, and this occurs (see Appendix) in the so-called *Arnold's tongues* (the very thin greenish regions). Increasing the coupling, the attractor undergoes a complex structure of bifurcations—not discussed in detail—that describe the classical torus-destruction route to chaos.· The genericity of Figure 15.3 confirms the robustness of the results.

Note that only a weak but not too small coupling allows feature (viii) to be matched, as the model attractor is periodic if the coupling is too strong, whereas the free-love model predicts quasi-periodic romantic regimes. Interestingly, chaos can be found for $\varepsilon = 0$, but not for $\delta = 0$, suggesting that the complaisance of Jules and the jealousy of Jim are the key elements triggering the unpredictability of the story.

The reader interested in checking the predictions of the K-J-J model (15.1–15.4) for different parameter values can do so interactively, using the online simulator described in Dercole and Rinaldi (2014).

15.5   The genius of Truffaut

A general message we can extract from this study is that a mathematical model can be used to highlight the genius of an artist—here François Truffaut, who featured "Jules et Jim" in his most important film, made in 1961, after discussing the idea with Roché.

Jeanne Moreau and Oskar Werner, already well known, played Kathe and Jules, while Henri Serre, selected because of a certain resemblance to Roché, played Jim. Truffaut omits many minor characters of the novel but successfully reproduces the feelings between Helen Grund and the two friends. Indeed Helen Grund, the only one of the three who could watch the film after Hessel and Roché passed away, wrote to Truffaut:

> *But what disposition in you, what affinity could have enlightened you to the point of recreating—in spite of the odd inevitable deviation and compromise—the essential quality of our intimate emotions?*

Truffaut actually adds, here and there, explicit elements pointing to the fact that love stories can be turbulent because of attracting and repelling phases that can be interpreted as folding and stretching phases, respectively. To show this, we use four segments (JJ1–JJ4) of the film available at

`home.deib.polimi.it/rinaldi/JulesEtJim`

Figure 15.5 shows still frames of the four segments and their initial and final times. The first of these segments anticipates the beginning of the

(JJ1) The initial lapidary message

(JJ2) Le turbillon de la vie

(JJ3) Bicycling in the countryside

(JJ4) Kathe and Jim burial

Fig. 15.5   Still frames of the four film segments (JJ1–JJ4) described in the text.

film. While the screen is still dark, Jeanne Moreau sends this succinct message:

Tu m'as dit: "Je t'aime."	*You told me: "I love you."*
Je t'ai dit: "Attends."	*I told you: "Wait."*
J'allais dire: "Prends-moi."	*I almost said: "Yes."*
Tu m'as dit: "Va-t'en."	*You said: "Go."*

Symmetrically, the film ends (see segment JJ4) by stressing again the antagonism between attraction and repulsion. After the suicide of Kathe and Jim, the voice-over says:

Les cendres furent recueillies dans des urnes et rangées dans un casier que l'on scella.	*The ashes were placed in an urn.*
Seul Jules les eût mêlées.	*Jules might have mixed them.*
Catherine avait toujours souhaité qu'on jetât les siennes dans le vent...	*Catherine wanted hers to be cast to the wind...*

Other stylistic elements that perfectly fit with chaotic dynamics are the use of the handy camera and of scenes in quick cuts, and the voice-over technique introduced by Truffaut to lump together long periods of calm and expand short critical periods like the moments when Kathe is ready to change partner. This is identical to what we do when we present strange attractors to students. For example, when we discuss the Lorenz butterfly attractor (see Figure A.5) we spend more time describing the rare and almost instantaneous switches from one wing to the other than the very long periods during which the trajectory rotates on the same wing.

But the most explicit reference to attraction and repulsion is *Le tourbillon de la vie* (the vortex of life), the soundtrack sung by Jeanne Moreau accompanied on the guitar by Albert, a friend of Kathe (see segment JJ2). This song is a beautiful hymn to chaos, characterized by recurrent phases of convergence and divergence:

On s'est connus, on s'est reconnus,	*We met with a kiss*
On s'est perdus d'vue, on s'est r'perdus d'vue	*A hit, then a miss*
On s'est retrouvés, on s'est réchauffés,	*It wasn't all bliss*
Puis on s'est séparés.	*And we parted*
Chacun pour soi est reparti.	*We went our own ways*
Dans l'tourbillon de la vie	*In life's whirlpool of days*
Je l'ai revue un soir, hàie, hàie, hàie	*I saw her again one night*
Ça fait déjà un fameux bail	*Again she was an enchanted sight*

And if that weren't, enough, Truffaut reinforces the message visually by introducing in the next scene (see segment JJ3) an effective representation of the divergence of nearby trajectories: Kathe, Jules, Jim, and Albert are bicycling in the countryside, when suddenly Albert leaves the group by taking a side road. This is analogous to what happens to four initially close trajectories in the Lorenz system: for some time the trajectories develop on the same wing of the attractor, but then one suddenly jumps on the other wing. A simulation of four initially close trajectories in the Lorenz system can be found at the end of segment JJ3.

All this was done by Truffaut in 1961, two years before the publication of Lorenz's famous paper (Lorenz, 1963). Thus, in our opinion, Truffaut grasped the role of stretching and folding in generating complex dynamics. In other words, he sensed, as only an artist can do, the basic mechanisms of an important scientific discovery and represented them with grace and skill.

Appendix A

Appendix

This Appendix is devoted to readers who are not yet familiar with mathematical modeling and, in particular, with dynamical systems and their bifurcations. The readers who have never worked with mathematical models could read the Appendix entirely, starting with the first section, where the sense of developing a model is explained through a simple example concerning the rate of sexual intercourse in a couple. In contrast, it is suggested that readers who already have some experience in mathematical modeling but are not familiar with the theory of dynamical systems and its jargon (trajectories, equilibria, limit cycles and their stability) start with Section A.2. Finally, those who only lack the notion of bifurcations and structural stability can start from Section A.3.

A.1 A simple illustrative example

In any field of science there are plenty of things which are known (from experience, from field observations, or from laboratory data) but which have not yet been satisfactorily explained. For example, everybody knew that prey and predator can have remarkable ups and downs or that epidemics developed from a few infectious individuals, reach a maximum, and then decline, but only Lotka, Volterra, Kermack, and McKendrick explained with their conceptual models why this is so, marking with these studies the beginning of modern ecology and epidemiology. This shows that knowing why something happens can in science be more important than knowing it just happens.

So far, romantic relationships have been studied only on the basis of data collected through interviews or self-reports or by psychoanalysts or therapists. This is rather surprising because in many other fields of science like, for example, economics and demography, conceptual deterministic models derived from general axioms (often conjectured without the use of data) are sometimes even more appreciated than empirical models based on data. Since there is no reason to believe that conceptual models cannot also be powerful descriptive tools in the study of romantic relationships, this book is entirely devoted to the presentation of this kind of model in the context of love dynamics.

For those who have never developed a conceptual model, we present in this section a very simple example taken from sexology in order to focus on the sense of using these models in science.

The time evolution of the rate of sexual intercourse has been discussed only empirically until recently (Rinaldi *et al.*, 2012), *i.e.*, through detailed statistical analysis of available data. From many of these studies it emerges that sexual activities decline over time in particular in the long term, but can sometimes increase in the first phase of the life of a couple. One can reasonably suspect that a property of this kind might be the mere consequence of the psycho-physiological behavioral characteristics of the partners. If so, they should be derivable on a purely logical basis (*i.e.*, without the use of data) from a few axioms describing the characteristics of the couple. This is what we show in this section by presenting a so-called conceptual model, namely, a set of axioms describing in mathematical terms the conjectured relationships between erotic potential, sexual appetite, and rate of sexual intercourse. For didactic reasons, we use a rather simple and scholastic style of presentation in order to show how a problem that has been investigated empirically until now can also be attacked from a purely dogmatic point of view.

Conceptual models are based on a number of simplifying assumptions that are often quite crude with respect to reality. In general, conceptual models involve two types of variables: stocks, namely, amounts of particular matters or properties, and flows, namely, rates of variation over time of matters or properties. For example, in models of the human body, stocks can be amounts of chemical compounds present at time τ in various organs, while flows represent the frequency at which these compounds are transferred from one organ to another or eliminated from the body. In the case of sexual intercourse, in order to minimize the dimensions of the model, *i.e.*, the number of variables and equations involved, our assumptions are actually extreme. In fact, we consider the couple as a compact unit with two stocks, called erotic potential $X(\tau)$ and sexual appetite $Y(\tau)$, and one flow, namely, the frequency of sexual intercourse $Z(\tau)$. The erotic potential

$X(\tau)$ is an internal resource characterizing a couple with life-time equal to τ. More precisely, a couple is initially (*i.e.*, at time $\tau = 0$) endowed with an erotic potential $X(0)$ (natural predisposition of the couple to sex), which is basically unknown to the two individuals, who have, however, the chance to gradually discover it through sexual intercourse. Each discovery is associated with a drop in the as yet undiscovered erotic potential $X(\tau)$ which is, therefore, systematically eroded by sex activity during the entire life of the couple. The sexual appetite of the couple $Y(\tau)$ is another stock variable representing in a compact way, *i.e.*, with a single positive variable, the desire for sex within the couple. This is perhaps the most questionable point of the model since sexual appetite is certainly simple in men, yet difficult to understand and recognize in women. Finally, we do not consider details on the quality and diversity of sexual activities, so that the frequency of sexual intercourse $Z(\tau)$ must simply be imagined as the number of standard sexual intercourse per unit time. Age and social environment are also not taken into account in order to obtain a time-invariant model, *i.e.*, a model with constant parameters. Under these assumptions, a conceptual model is nothing but a set of conjectures (often called axioms) on the relationships existing among the variables $X(\tau)$, $Y(\tau)$, and $Z(\tau)$. Of course, there are many possible conjectures, and some of them may be *a priori* more reasonable than others in particular contexts, for example, when applied to couples from specific cultures. Here, we claim only to give an example of conceptual models and therefore focus on three very simple conjectures:

(1) The discovery of erotic potential associated with a single act of intercourse is proportional to the erotic potential itself. Thus, the discovery of erotic potential in an interval of length Δ characterized by a given number of acts of intercourse is proportional to that number times the erotic potential. This allows one to write the following balance equation over any interval starting at a generic time τ and ending at time $\tau + \Delta$

$$X(\tau + \Delta) = X(\tau) - \alpha X(\tau) Z(\tau) \Delta. \tag{A.1}$$

In this equation $Z(\tau)\Delta$ is a fair approximation of the number of acts of intercourse during the interval $[\tau, \tau + \Delta]$ if Δ is so small (*e.g.*, one month, in the case at hand) that the frequency of intercourse remains practically constant during the interval. If $Z(\tau)$ is known for a series of successive values of τ, say $\tau = 0, \Delta, 2\Delta, 3\Delta, \ldots$, the above equation can be used recursively to produce the series of erotic potential $X(\Delta), X(2\Delta), X(3\Delta), \ldots$ once the initial endowment $X(0)$ of erotic potential is known. Discrete-time models of this kind are particularly useful for performing simulations.

(2) Sexual appetite is reduced by sexual activity and, at the same time, regenerated by the expectation of future pleasure which, in turn, is positively correlated with discovered erotic potential. More precisely, each act of intercourse reduces sexual appetite by a fixed amount γ, while each unit of discovered erotic potential regenerates sexual appetite of a fixed amount β. Thus, the balance equation over an interval $[\tau, \tau + \Delta]$ is

$$Y(\tau + \Delta) = Y(\tau) + \beta(X(\tau) - X(\tau + \Delta)) - \gamma Z(\tau)\Delta. \qquad (A.2)$$

(3) The frequency of sexual intercourse is proportional to sexual appetite, *i.e.*,

$$Z(\tau) = \delta Y(\tau).$$

This equation is not a recursive equation as (A.1) and (A.2) and allows sexual appetite to be identified with frequency of sexual intercourse (module a proportionality factor). By substituting $Y(\tau)$ with $Z(\tau)/\delta$ in equation (A.2) one obtains

$$Z(\tau + \Delta) = Z(\tau) - \beta\delta(X(\tau + \Delta) - X(\tau)) - \gamma\delta Z(\tau)\Delta, \qquad (A.3)$$

which, together with equation (A.1), is the discrete time model formally interpreting our three axioms.

Discrete-time models of the kind (A.1,A.3) are not the most standard description of dynamic processes. In fact in biology, economics, and physics, continuous-time models are usually preferred. These models are nothing but discrete-time models with extremely small time intervals Δ. More precisely, in the case under consideration, the continuous-time model can be derived in two steps as follows. First divide equations (A.1,A.3) by Δ, thus obtaining

$$\frac{X(\tau + \Delta) - X(\tau)}{\Delta} = -\alpha X(\tau)Z(\tau)$$
$$\frac{Z(\tau + \Delta) - Z(\tau)}{\Delta} = -\beta\delta\frac{X(\tau + \Delta) - X(\tau)}{\Delta} - \gamma\delta Z(\tau) \qquad (A.4)$$

and then let Δ tend to zero. But the limits of the left-hand sides of these two equations are, by definition, the so-called time-derivatives of $X(\tau)$ and $Z(\tau)$, universally indicated with $dX(\tau)/d\tau$ and $dZ(\tau)/d\tau$ or, more simply, with $\dot{X}(\tau)$ and $\dot{Z}(\tau)$. Thus, the continuous-time model is composed of the following two equations, called ordinary differential equations,

$$\dot{X}(\tau) = -aX(\tau)Z(\tau)$$
$$\dot{Z}(\tau) = bX(\tau)Z(\tau) - cZ(\tau) \qquad (A.5)$$

where $a = \alpha$, $b = \alpha\beta\delta$, and $c = \gamma\delta$.

The solutions to equations (A.5) for any given initial condition $(X(0), Z(0))$ cannot be obtained in closed form, but they can be derived numerically and represented as in Figure A.1a or with a single curve in the space (X, Z) as shown in Figure A.1b. Each curve in the space (X, Z) develops in time from the right to the left since $X(\tau)$ is always decreasing (its derivative $\dot{X}(\tau)$ is always negative, *i.e.*, the erotic potential can only be eroded through sex), while $Z(\tau)$ first increases if

$$X(0) > \frac{c}{b} \tag{A.6}$$

and then decreases. The frequency of sexual intercourse is maximum when sexual appetite is maximum $(\dot{Y}(\tau) = 0)$, *i.e.*, when the erotic potential is equal to the characteristic value $c/b = \gamma/(\alpha\beta)$. Thus, if the couple is initially endowed with too low an erotic potential $X(0)$, *i.e.*, if (A.6) is not satisfied, then the sexual activity declines from the start, while in the opposite case it first increases and then declines (see Figure A.1). As noted

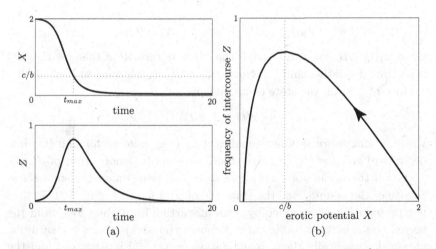

Fig. A.1 Time evolution of erotic potential (X) and frequency of sexual intercourse (Z) predicted by model (A.5) for $a = c = 1$ and $b = 2$. The frequency of sexual intercourse peaks at t_{max}, where $X(t_{max}) = c/b$.

earlier, both possibilities are confirmed by data. All curves in Figure A.1 tend asymptotically toward the segment $(0, c/b)$ of the X axis, *i.e.*, the frequency of intercourse tends to zero in the long term, and the residual erotic potential is small.

A.2 Dynamical systems and state portraits

In this section we summarize the basic definitions and tools of analysis of dynamical systems, with particular emphasis on the asymptotic behavior of continuous-time systems described by Ordinary Differential Equations (ODEs). The literature on dynamical systems is huge and we do not attempt to survey it here. Most of the results on continuous-time systems are described in more detail in Kuznetsov (2004), while less formal but didactically very effective treatments, rich in interesting examples and applications, can be found in Strogatz (1994) and Alligood *et al.* (1996).

The dynamical systems considered in this chapter are *continuous-time, finite-dimensional* dynamical systems described by n *autonomous* (*i.e.,* time-independent) ODEs called *state equations, i.e.,*

$$\dot{x}_1(t) = f_1(x_1(t), x_2(t), \ldots, x_n(t)),$$
$$\dot{x}_2(t) = f_2(x_1(t), x_2(t), \ldots, x_n(t)),$$
$$\vdots$$
$$\dot{x}_n(t) = f_n(x_1(t), x_2(t), \ldots, x_n(t)),$$

where $x_i(t) \in \mathbf{R}$, $i = 1, 2, \ldots, n$, is the ith *state variable* at time $t \in \mathbf{R}$, $\dot{x}_i(t)$ is its time derivative, and functions $f_1, \ldots, f_n$ are assumed to be smooth.

In vector form, the state equations are

$$\dot{x}(t) = f(x(t)), \tag{A.7}$$

where x and $\dot{x}$ are n-dimensional vectors (the *state vector* and its time derivative) and $f = [f_1, \ldots, f_n]^T$ (the T superscript denotes transposition).

Given the initial state $x(0)$, the state equations uniquely define a *trajectory* of the system, *i.e.,* the state vector $x(t)$ for all $t \geq 0$. A trajectory is represented in state space by a curve starting from point $x(0)$, and the vector $\dot{x}(t)$ is tangent to the curve at point $x(t)$. Trajectories can easily be obtained numerically through simulation (numerical integration) and the set of all trajectories (one for any $x(0)$) is called *state portrait*. If $n = 2$ (*second-order* or *planar* systems) the state portrait is often represented by drawing a sort of qualitative skeleton, *i.e.,* strategic trajectories (or finite segments of them), from which all other trajectories can be intuitively inferred. For example, in Figure A.2a the skeleton is composed of 13 trajectories: three of them (A, B, C) are just points (corresponding to constant solutions of (A.7)) and are called *equilibria*, while one (γ) is a closed trajectory (corresponding to a periodic solution of (A.7)) called *limit cycle*. The

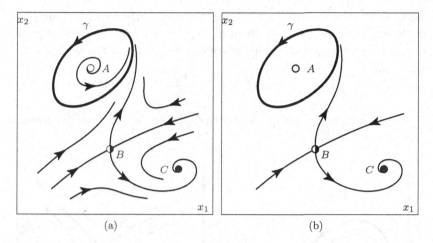

Fig. A.2   Skeleton of the state portrait of a second-order system: (a) skeleton with 13 trajectories; (b) reduced skeleton (characteristic frame) with 8 trajectories (attractors, repellers, and saddles with stable and unstable manifolds).

other trajectories allow one to conclude that A is a *repeller* (no trajectory starting close to A tends or remains close to A), B is a *saddle* (almost all trajectories starting close to B go away from B, but two trajectories tend to B and compose the so-called *stable manifold*; the two trajectories emanating from B compose the *unstable manifold* and both manifolds are also called *saddle separatrices*), while C and γ are *attractors* (all trajectories starting close to C [γ] tend to C [γ]). Attractors are *asymptotically stable* if they attract all nearby trajectories and *globally stable* if they attract all trajectories (technically with the exclusion of sets with no measure in state space), while saddles and repellers are *unstable*.

The skeleton of Figure A.2a also identifies the *basin of attraction* of each attractor: in fact, all trajectories starting above [below] the stable manifold of the saddle tend toward the limit cycle γ [the equilibrium C]. Note that the basins of attraction are open sets as their boundaries are the saddle and its stable manifold. Often, the full state portrait can be more easily imagined when the skeleton is reduced, as in Figure A.2b, to its basic elements, namely, attractors, repellers, and saddles with their stable and unstable manifolds. From now on, the reduced skeleton is called the *characteristic frame*.

Limit cycles are the elements of the characteristic frame that are most difficult to detect. For this reason, conditions for their existence (or non-

existence) and uniqueness are particularly useful. Here we recall only one
of these conditions, namely, *Bendixon's criterion*, which says that cycles
cannot exist in bounded regions where the divergence of the system

$$\mathrm{div} f = \frac{\partial f_1}{\partial x_1} + \frac{\partial f_2}{\partial x_2},$$

does not change sign and is, possibly, equal to zero on one-dimensional
curves.

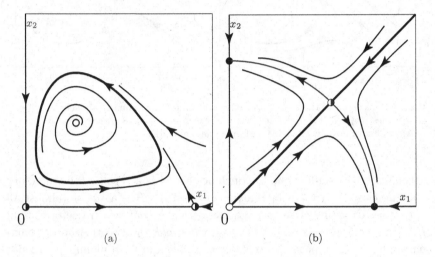

Fig. A.3   Skeleton of the state portrait of two populations: (a) cyclic coexistence of prey
and predator; (b) competitive exclusion.

State portraits of second-order systems have been the main tool of communication in many fields of science. For example, the two state portraits in
Figure A.3 represent the first important discoveries on predation and competition, respectively. Figure A.3a, where x_1 and x_2 are prey and predator,
shows that the two populations evolve toward a periodic behavior for all
generic initial conditions (cyclic coexistence), while Figure A.3b, where x_1
and x_2 compete for the same resource, shows that only one of the two
competitors remains in the game in the long term (principle of competitive
exclusion).

The asymptotic behaviors of continuous-time second-order systems are
quite simple, because in the case $n = 2$ attractors can be equilibria (*stationary regimes*) or limit cycles (*cyclic* or *periodic regimes*). But in higher-
dimensional systems, *i.e.*, for $n \geq 3$, more complex behaviors are possible

since attractors can also be *tori* (*quasi-periodic regimes*) or *strange attractors* (*chaotic regimes*).

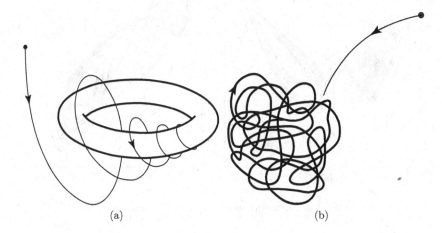

(a)   (b)

Fig. A.4   Sketch of an attracting torus (a) and of a strange attractor (b).

A torus attracting nearby trajectories is sketched in Figure A.4a. A trajectory starting from a point of the torus remains forever on it (*i.e.*, the torus is *invariant* for the dynamics of the system) but, in general, never passes again through the starting point. For example, two frequencies characterize a three-dimensional torus, namely, two positive real numbers, $1/T_1$, $1/T_2$, measuring the number of rotations around the cross-section of the torus and the number of revolutions along it, per unit of time. Since, generically, the ratio T_1/T_2 is irrational there is no period T such that

$$T = T_1 r_1 = T_2 r_2, \tag{A.8}$$

with r_1 and r_2 positive integers. In words, there is no time T in which a trajectory on the torus carries out an integer number of cross-section rotations and an integer, possibly different, number of torus revolutions, *i.e.*, no time T after which the trajectory revisits the starting point. As a consequence, a single trajectory on the torus covers it densely in the long term, and the corresponding regime is called quasi-periodic, being the result of two (or more in higher-dimensions) frequencies.

In special cases, however, the ratio T_1/T_2 can be rational, *i.e.*, trajectories on torus can be periodic (($r_1 : r_2$) *cycles on torus*, for the minimum r_1 and r_2 satisfying (A.8)) A cycle on torus can be stable (*i.e.*, attracting nearby trajectories on torus) or unstable. For obvious

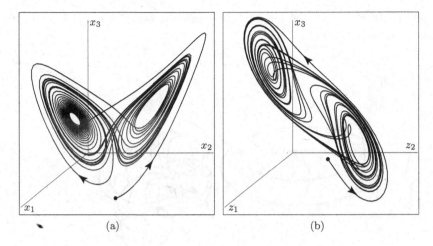

Fig. A.5   The Lorenz (a) and the Chua (b) strange attractors.

topological reasons, the existence of a stable $(r_1:r_2)$ cycle on torus requires the existence of an unstable $(r_1:r_2)$ cycle on the same torus, and rules out cycles characterized by different pairs (r_1, r_2).

A strange attractor (a sort of "tangle" in state space) is shown in Figure A.4b. Trajectories starting in the vicinity of the tangle tend to it and then remain in it forever. Strange attractors generated by mathematical models are often quite regular, as pointed out in Figure A.5 where two famous strange attractors are drawn: the butterfly Lorenz attractor (a) and the double-scroll Chua's attractor (b).

The most striking difference among attractors is that equilibria, cycles, and tori have integer dimension (0, 1, and 2, respectively), while strange attractors are *fractal* sets and therefore have noninteger dimension (see next section). Another important difference is that two trajectories starting from very close points in the same basin of attraction remain very close forever if the attractor is an equilibrium, a cycle, or a torus, while they alternatively diverge (*stretching*) and converge (*folding*) forever if the attractor is a tangle. The mean rate of divergence of nearby trajectories is measured by the so-called *Lyapunov exponent*, which is the most important indicator in the study of deterministic chaos.

In the simple but very important case of linear systems

$$\dot{x}(t) = Ax(t),$$

the state portrait can immediately be obtained from the eigenvalues and eigenvectors of the $n \times n$ matrix A. We recall that the eigenvalues of an

$n \times n$ matrix A are the zeros $\lambda_1, \lambda_2, \ldots, \lambda_n$ of its characteristic polynomial $\det(\lambda I - A)$, where det denotes matrix determinant, and that the eigenvectors associated with an eigenvalue λ_i are nontrivial vectors $x^{(i)}$ satisfying the relationship $Ax^{(i)} = \lambda_i x^{(i)}$. There are five generic state portraits of

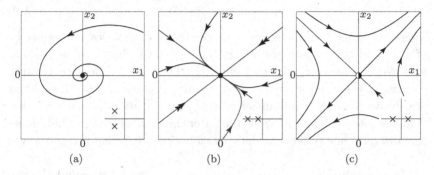

(a)   (b)   (c)

Fig. A.6 Three state portraits of generic second-order continuous-time linear systems ($\lambda_1 \neq \lambda_2$, both with non-zero real part, see the complex plane associated with each panel): (a) (stable focus) and (b) (stable node) are attractors; the unstable focus (positive real part complex conjugate eigenvalues) and the unstable node (positive real eigenvalues) (repellers) are obtained by reversing all arrows in the state portraits (a) and (b), respectively; (c) is a saddle. Straight trajectories correspond to eigenvectors associated with real eigenvalues. Double arrows indicate the straight trajectories along which the state varies more rapidly.

second-order continuous-time linear systems: three of them are shown in Figure A.6 (the other two are obtained from cases (a) and (b) by reversing the sign of the eigenvalues and all arrows in the state portraits).

When the two eigenvalues are complex (case a), the trajectories spiral around the origin and tend to [diverge from] it if the real part of the eigenvalues is negative [positive]. By contrast, when the two eigenvalues are real (cases (b) and (c)), the trajectories do not spiral and there are actually special straight trajectories (corresponding to the eigenvectors) converging to [diverging from] the origin if the corresponding eigenvalue is negative [positive]. Along the straight trajectories both state variables vary in time as $\exp(\lambda_i t)$, while along all other trajectories they follow a more complex law of the kind $c_1 \exp(\lambda_1 t) + c_2 \exp(\lambda_2 t)$. This is because in generic cases $\lambda_1 \neq \lambda_2$, one of the two exponential functions dominates the other for $t \to \pm\infty$ and all curved trajectories tend to align with one of the two straight trajectories. In particular, in the case of a stable node (characterized by $\lambda_2 < \lambda_1 < 0$, see Figure A.6b), both exponential functions tend to zero for $t \to +\infty$, but in the long term $\exp(\lambda_1 t) \gg \exp(\lambda_2 t)$ so that all trajectories, except the two straight trajectories corresponding to the second eigenvector $x^{(2)}$, tend to zero tangentially to the first eigenvector $x^{(1)}$.

Very similar definitions can be given for *discrete-time systems* described by n difference equations of the form

$$x(t+1) = f(x(t)), \qquad (A.9)$$

where the time t is an integer. In this case trajectories are sequences of points in state space and, again, asymptotic regimes can be stationary, cyclic, quasi-periodic, and chaotic. The major difference between continuous-time and discrete-time dynamical systems is that the former are always *reversible*, since under very general conditions system (A.7) has a unique solution for $t < 0$, while the latter can be *irreversible*. This implies that discrete-time systems can have quasi-periodic and chaotic regimes even if $n = 1$.

The equilibria of system (A.7) can be found by determining all solutions $\bar{x}$ of (A.7) with $\dot{x} = 0$. In second-order systems the equilibria are often determined graphically through the so-called *null-clines*, which are nothing but the lines in state space on which $f_1(x_1, x_2) = 0$ (x_1-null-clines) and $f_2(x_1, x_2) = 0$ (x_2-null-clines). Obviously, the equilibria are at the intersections of x_1- and x_2-null-clines. Moreover, all trajectories cross x_1- [x_2-] isoclines vertically [horizontally] because $\dot{x}_1$ [$\dot{x}_2$] is zero on x_1- [x_2-] null-clines. This property is often used for devising qualitative geometric features of the state portrait. Null-clines are also useful in the discussion of the influence of a parameter in the dynamics of the system. This is particulary true when the parameter influences only one of the two null-clines, as in most cases analyzed in this book.

The stability of an equilibrium $\bar{x}$ is not as easy to ascertain. However, it can very often be discussed through *linearization*, *i.e.*, by approximating the behavior of the system in the vicinity of the equilibrium through a linear system. This can be done in the following way. Let

$$\delta x(t) = x(t) - \bar{x},$$

so that

$$\dot{\delta x}(t) = f(\bar{x} + \delta x(t)).$$

Under very general conditions, we can expand the function f in Taylor series, thus obtaining

$$\dot{\delta x}(t) = f(\bar{x}) + \left. \frac{\partial f}{\partial x} \right|_{x=\bar{x}} \delta x(t) + O(\|\delta x(t)\|^2),$$

where $\|\cdot\|$ is the standard norm in $\mathbf{R}^n$ and $O(\|\delta x(t)\|^2)$ stays for a term that vanishes as $\|\delta x(t)\|^2$ when $\delta x(t) \to 0$. Noting that $f(\bar{x}) = 0$, since $\bar{x}$ is a constant solution of (A.7), we have

$$\dot{\delta x}(t) = \left.\frac{\partial f}{\partial x}\right|_{x=\bar{x}} \delta x(t) + O(\|\delta x(t)\|^2), \qquad (A.10)$$

where the $n \times n$ constant matrix

$$J = \left.\frac{\partial f}{\partial x}\right|_{x=\bar{x}} = \begin{bmatrix} \dfrac{\partial f_1}{\partial x_1} & \cdots & \dfrac{\partial f_1}{\partial x_n} \\ \vdots & & \vdots \\ \dfrac{\partial f_n}{\partial x_1} & \cdots & \dfrac{\partial f_n}{\partial x_n} \end{bmatrix}_{x=\bar{x}} \qquad (A.11)$$

is called the *Jacobian matrix* (or, more simply, *Jacobian*). One can easily imagine that, under suitable conditions, the behavior of system (A.10) (which is still system (A.7)) can be well approximated in the vicinity of $\bar{x}$, by the so-called *linearized system*, which, by definition, is

$$\dot{\delta x}(t) = \left.\frac{\partial f}{\partial x}\right|_{x=\bar{x}} \delta x(t). \qquad (A.12)$$

This is indeed the case. In particular, it can be shown that if the solution $\delta x(t)$ of (A.12) tends to 0 for all $\delta x(0) \neq 0$ (as in Figures A.6a and A.6b), then the same is true for system (A.10) provided that $\|\delta x(0)\|$ is sufficiently small. In other words, the stability of the linearized system implies the (local) stability of the equilibrium $\bar{x}$. This result is quite interesting because the stability of the linearized system can be numerically ascertained by checking if all eigenvalues λ_i, $i = 1, \ldots, n$, of the Jacobian matrix (A.11) have negative real part. A similar result also holds for the case of unstable equilibria. More precisely, if at least one eigenvalue λ_i of the Jacobian matrix has positive real part (as in Figure A.6c), then the equilibrium $\bar{x}$ is locally unstable (*i.e.*, the solution of (A.10) diverges at least temporarily from zero for suitable $\delta x(0)$, no matter how small $\|\delta x(0)\|$ is).

Similarly, the local stability of an equilibrium of a discrete-time system of the form (A.9) can be studied by simply looking at the module $|\lambda_i|$ of the n eigenvalues λ_i. In fact, if all $|\lambda_i| < 1$, *i.e.*, if all eigenvalues are inside the unit circle in the complex plane, the equilibrium is stable, while if at least one eigenvalue is outside the unit circle ($|\lambda_i| > 1$), the equilibrium is unstable.

The study of the stability of limit cycles can also be carried out through linearization, following a very simple idea suggested by Poincaré (see Figure A.7). In the case of second-order systems (see Figure A.7a) the Poincaré

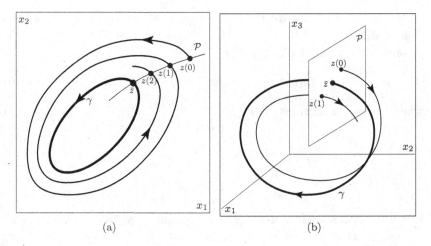

Fig. A.7 A stable limit cycle γ, the Poincaré section $\mathcal{P}$, and the sequence $z(0), z(1), z(2), \ldots$ of return points.

method consists of cutting locally and transversally the limit cycle with a manifold $\mathcal{P}$, called the *Poincaré section*, and looking at the sequence $z(0), z(1), z(2), \ldots$ of points of return of the trajectory to $\mathcal{P}$. Since $\mathcal{P}$ is one-dimensional, $z(t)$ is a scalar coordinate on $\mathcal{P}$ and the state equation (A.7) implicitly defines a first-order discrete-time system called the *Poincaré map*

$$z(t+1) = P(z(t)). \tag{A.13}$$

The intersection $\bar{z}$ of the limit cycle γ with $\mathcal{P}$ is an equilibrium of the Poincaré map (since $\bar{z} = P(\bar{z})$) and γ is stable if and only if the equilibrium $\bar{z}$ of (A.13) is stable. One can therefore use the linearization technique, by taking into account that the eigenvalue of the linearized Poincaré map, $dP/dz|_{z=\bar{z}}$ (called the *Floquet multiplier*, or simply multiplier, of the cycle), cannot be negative, since trajectories cannot cross each other. Thus, a sufficient condition for the (local) stability of the limit cycle γ is

$$\left. \frac{dP}{dz} \right|_{z=\bar{z}} < 1, \tag{A.14}$$

while the reverse inequality implies the instability of γ.

Similarly, in the case of third-order systems (see Figure A.7b) the Poincaré section is a two-dimensional manifold $\mathcal{P}$ and the points of return $z(0), z(1), z(2), \ldots$ are generated by a two-dimensional Poincaré map (A.13). Again, the cycle is stable if and only if the equilibrium $\bar{z}$ of the discrete-time system (A.13) is stable. Thus, if the two multipliers of the cycle, *i.e.*, the two eigenvalues of the Jacobian matrix $\partial P/\partial z|_{z=\bar{z}}$, are smaller than 1 in module, the cycle γ is stable, while if the module of at least one multiplier is greater than 1 the cycle is unstable. These sufficient conditions for the stability and instability of a cycle can obviously be extended to the n-dimensional case, where $\partial P/\partial z|_{z=\bar{z}}$ is an $(n-1) \times (n-1)$ matrix. It must be noted, however, that they can only be verified numerically, since the cycle γ is in general not known analytically.

The Poincaré section is also very useful for distinguishing quasi-periodic from chaotic regimes in third-order systems. In fact, a torus appears on a Poincaré section as a regular closed curve, while strange attractors appear as fractal sets (clouds of points), as shown in Figure A.8.

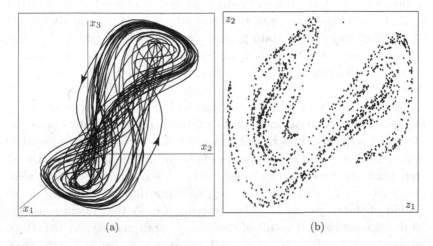

(a)   (b)

Fig. A.8   Strange attractor of the three-dimensional Ueda system: (a) in state space; (b) on a Poincaré section.

A.3   Structural stability

In this and the following sections we discuss the possible structural changes of the asymptotic behavior of dynamical systems under parameter variation, called *bifurcations*. The literature on bifurcation analysis is huge and we do not attempt to survey it here. Most of the results on bifurcations of dynamical systems are described in detail in Kuznetsov (2004).

Structural stability is a key notion in the theory of dynamical systems, as it is needed to understand interesting phenomena like catastrophic transitions, bistability, hysteresis, frequency locking, synchronization, subharmonics, deterministic chaos, and others. The final target of structural stability is the study of the asymptotic behavior of parameterized families of dynamical systems of the form

$$\dot{x}(t) = f(x(t), p), \tag{A.15}$$

for continuous-time systems, and

$$x(t+1) = f(x(t), p), \tag{A.16}$$

for discrete-time systems, where p is a vector of constant *parameters*. Given the parameter vector p, all the definitions that we have seen in the previous section apply to the particular dynamical system of the family identified by p. Thus, all geometric and analytical properties of systems (A.15) and (A.16), *e.g.*, trajectories, state portrait, equilibria, limit cycles, their stability and associated Jacobian matrices and Poincaré maps, the basins of attraction, and, consequently, the asymptotic behavior of the system, now depend on p.

Structural stability allows one to rigorously explain why a small change in a parameter value can give rise to a radical change in system behavior. More precisely, the aim is to find regions $\mathcal{P}_i$ in parameter space characterized by the same qualitative behavior of system (A.15), in the sense that all state portraits corresponding to values $p \in \mathcal{P}_i$ are topologically equivalent (*i.e.*, they can be obtained one from the other through a smooth deformation of the trajectories). Thus, varying $p \in \mathcal{P}_i$ the system conserves all the characteristic elements of the state portrait, namely, its attractors, repellers, and saddles. In other words, when p is varied in $\mathcal{P}_i$, the characteristic frame varies but conserves its structure. Figure A.9 shows the typical result of a study of structural stability in the space (p_1, p_2) of two parameters of a second-order system. The parameter space is subdivided into three regions, $\mathcal{P}_1$, $\mathcal{P}_2$, and $\mathcal{P}_3$, and for all interior points of each one of these regions the state portrait is topologically equivalent to that sketched in the figure. In $\mathcal{P}_1$ the system is an oscillator, since it has a single attractor which is a limit cycle. Also in $\mathcal{P}_2$ there is a single attractor, which is, however, an equilibrium. Finally, in $\mathcal{P}_3$ we have *bistability* since the system has two alternative attractors (two equilibria), each with its own basin of attraction delimited by the stable manifold of the saddle equilibrium.

If p is an interior point of a region $\mathcal{P}_i$, system (A.15) is said to be *structurally stable* at p since its state portrait is qualitatively the same as those of the systems obtained by slightly perturbing the parameters in all possible ways. By contrast, if p is on the boundary of a region $\mathcal{P}_i$ the system

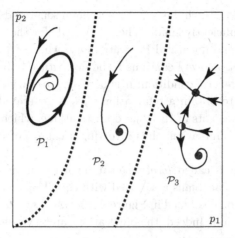

Fig. A.9   Bifurcation diagram of a second-order system. The curves separating regions $\mathcal{P}_1$, $\mathcal{P}_2$, and $\mathcal{P}_3$ are bifurcation curves.

is not structurally stable as small perturbations can give rise to qualitatively different state portraits. The points of the boundaries of the regions $\mathcal{P}_i$ are called *bifurcation points*, and, in the case of two parameters, the boundaries are called *bifurcation curves*. Bifurcation points are therefore points of degeneracy. If they lie on a curve separating two distinct regions $\mathcal{P}_i$ and $\mathcal{P}_j$, $i \neq j$, they are called codimension-1 bifurcation points, while if they lie on the boundaries of three distinct regions they are called codimension-2 bifurcation points, and so on.

In the following, we mainly deal with second-order, continuous-time systems and focus on codimension-1 bifurcations.

A.4   Bifurcations as collisions

A generic element of the parameterized family of dynamical systems (A.15) must be imagined to be structurally stable because if p is selected randomly it will be an interior point of a region $\mathcal{P}_i$ with probability 1. In generic conditions, attractors, repellers, saddles, and their stable and unstable manifolds are separated one from another. Moreover, the eigenvalues of the Jacobian matrices associated with equilibria have non-zero real parts, while the eigenvalues of linearized Poincaré maps associated with cycles have module different from 1. By continuity, small parametric variations will induce small variations of all attractors, repellers, saddles, and their stable and unstable manifolds which, however, will remain separated if the parametric variations are sufficiently small. The same holds for the eigenvalues of Jacobian matrices and linearized Poincaré maps, which, for sufficiently small parametric variations, will continue to be noncritical. Thus, in conclusion, starting from a generic condition, it is necessary to vary the parameters of a finite amount to obtain a bifurcation; this is generated by the collision of two or more elements of the characteristic frame, which then changes its structure at the bifurcation, thus involving a change of the state portrait of the system.

A bifurcation is called *local* when it involves the degeneracy of some eigenvalue of the Jacobians associated with equilibria or cycles. For example, the bifurcation described in Figure A.10, called *saddle-node bifurcation*, is a local bifurcation. Indeed, the bifurcation can be viewed as the collision,

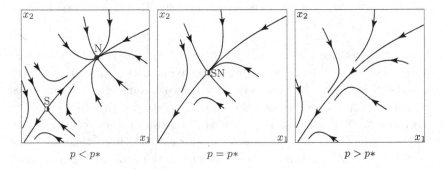

Fig. A.10   Example of local bifurcation: saddle-node bifurcation.

at $p = p^*$, of two equilibria: for $p < p^*$ the two equilibria (elements of the characteristic frame) are distinct and one is stable (the node N) while the other is unstable (the saddle S). Then, as p increases, the two equilibria approach each other and finally collide when $p = p^*$ and then disappear. Note that the characteristic frame is degenerate at $p = p^*$ because it is composed of one element (an equilibrium), while there are two equilibria for $p < p^*$ and none for $p > p^*$. But the bifurcation can also be interpreted in terms of eigenvalue degeneracy. In fact, the eigenvalues of the Jacobian evaluated at the saddle are one positive and one negative, while the eigenvalues of the Jacobian evaluated at the node are both negative, so that when the two equilibria coincide, one of the two eigenvalues of the unique Jacobian matrix must be equal to zero.

In contrast, *global bifurcations* cannot be revealed by eigenvalue degeneracies. One example, known as *heteroclinic bifurcation*, is shown in Figure A.11, which presents the characteristic frames (two saddles and their stable and unstable manifolds) of a system for $p = p^*$ (bifurcation value) and for $p \neq p^*$. The characteristic frame for $p = p^*$ is structurally differ-

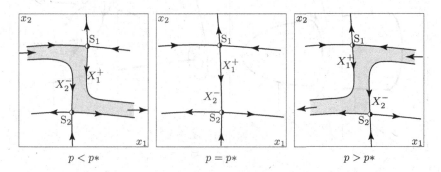

Fig. A.11   Example of global bifurcation: heteroclinic bifurcation.

ent from the others because it corresponds to the collision of the unstable manifold X_1^+ of the first saddle with the stable manifold X_2^- of the second saddle. However, the two Jacobian matrices associated with the two saddles do not degenerate at p^*, since their eigenvalues remain different from zero. In other words, the bifurcation cannot be revealed by the behavior of the system in the vicinity of an equilibrium, but is the result of the global behavior of the system.

When there is only one parameter p and there are various bifurcations at different values of the parameter, it is often advantageous to represent the dependence of the system behavior upon the parameter by drawing in the three-dimensional space (p, x_1, x_2), often called *control space*, the characteristic frame for all values of p. This is done, for example, in Figure A.12 for the same system described in Figure A.9, with $p = p_1$ and constant p_2. Figure A.12 shows that for increasing values of p a so-called *Hopf bifurcation* occurs, as the stable limit cycle shrinks to a point, thus colliding with the unstable equilibrium that exists inside the cycle. This is a local bifurcation, because the equilibrium is stable for higher values of p, so that the bifurcation can be revealed by an eigenvalue degeneracy. The figure also shows that a saddle-node bifurcation occurs at a higher value of the parameter, as two equilibria, namely, a stable node and a saddle, become closer and closer, when p decreases, until they collide and disappear. The Hopf and the saddle-node bifurcations are perhaps the most popular local

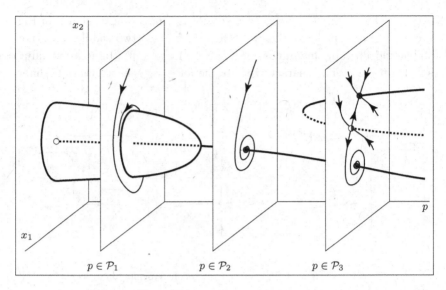

Fig. A.12 Characteristic frame in the control space of a system with a Hopf and a saddle-node bifurcation. Continuous lines represent trajectories in the three illustrated state portraits and stable equilibria or limit cycles otherwise; dashed lines represent unstable equilibria. The symbols $\mathcal{P}_1$, $\mathcal{P}_2$, and $\mathcal{P}_3$ refer to Figure A.9.

bifurcations of second-order systems and are discussed in some detail in the next section.

A.5   Local bifurcations

We now discuss the seven most important local bifurcations of continuous-time systems. Three of them, called *transcritical*, saddle-node (already encountered above), and *pitchfork*, can be viewed as collisions of equilibria. Since they can occur in first-order systems, we present them in that context. The other bifurcations involve limit cycles. Two of them can occur in second-order systems, namely, the Hopf bifurcation (already seen), *i.e.*, the collision of an equilibrium with a vanishing cycle, and the *tangent of limit cycles*, which is the collision of two cycles. The last two bifurcations, the *flip* (or *period-doubling*) and the *Neimark-Sacker* (or *torus*), are more complex because they can occur only in three- (or higher-) dimensional systems. The first is a particular collision of two limit cycles, one with a period that is double the other, while the second is the collision between a cycle and a vanishing torus.

Transcritical, saddle-node, and pitchfork bifurcations

Figure A.13 shows three different types of collisions of equilibria in first-order systems of the form (A.15). The state x and the parameter p have

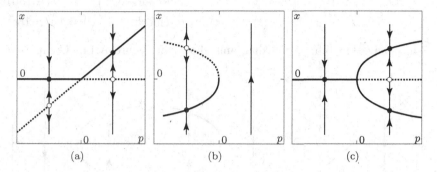

(a) (b) (c)

Fig. A.13 Three local bifurcations viewed as collisions of equilibria: (a) transcritical; (b) saddle-node; (c) pitchfork.

been normalized in such a way that the bifurcation occurs at $p^* = 0$ and that the corresponding equilibrium is zero. Continuous lines in the figure represent stable equilibria, while dashed lines indicate unstable equilibria. In Figure A.13a the collision is visible in both directions, while in Figures A.13b and A.13c the collision is visible only from the left or from the right. The three bifurcations are called, respectively, transcritical, saddle-node, and pitchfork, and the three most simple state equations (called *normal forms*) giving rise to Figure A.13 are

$$\dot{x}(t) = px(t) - x^2(t), \quad \text{transcritical}, \tag{A.17a}$$
$$\dot{x}(t) = p + x^2(t), \quad \text{saddle-node}, \tag{A.17b}$$
$$\dot{x}(t) = px(t) - x^3(t), \quad \text{pitchfork}. \tag{A.17c}$$

The first of these bifurcations is also called *exchange of stability* since the two equilibria exchange their stability at the bifurcation. The second is called saddle-node bifurcation because in second-order systems it corresponds to the collision of a saddle with a node, as shown in Figure A.10, but it is also known as *fold*, in view of the form of the graph of its equilibria. Due to the symmetry of the normal form, the pitchfork has three colliding equilibria, two stable and one unstable in the middle.

It is worth noting that if the signs of the quadratic and cubic terms in (A.17) are changed, three new normal forms are obtained, namely,

$$\dot{x}(t) = px(t) + x^2(t), \quad \text{transcritical}, \qquad \text{(A.18a)}$$

$$\dot{x}(t) = p - x^2(t), \quad \text{saddle-node}, \qquad \text{(A.18b)}$$

$$\dot{x}(t) = px(t) + x^3(t), \quad \text{pitchfork}, \qquad \text{(A.18c)}$$

which have the bifurcation diagrams shown in Figure A.14. Comparing

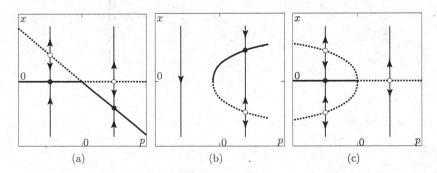

Fig. A.14   Bifurcation diagrams corresponding to the normal forms (A.18).

this figure with Figure A.14, it is easy to verify that nothing changes from a phenomenological point of view in the first two cases. However, for the pitchfork bifurcation this is not true, since in case (A.17c) there is at least one attractor for each value of the parameter, while in case (A.18c), for $p > 0$, there is only a repeller. To distinguish the two possibilities, the pitchfork (A.17c) is called *supercritical*, while the other is called *subcritical*.

Hopf bifurcation

The Hopf bifurcation (actually discovered by A. A. Andronov for second-order systems; see Andronov *et al.*, 1973, and Marsden and McCracken, 1976, for the English translation of Andronov and Hopf's original works) explains how a stationary regime can become cyclical as a consequence of a small variation of a parameter, a rather common phenomenon not only in physics but also in biology, economics, and life sciences. In terms of collisions, this bifurcation involves an equilibrium and a cycle which, however, shrinks to a point when the collision occurs. Figure A.15 shows the two possible cases, known as supercritical and subcritical Hopf bifurcations, respectively. In the supercritical case, a stable cycle has in its interior an

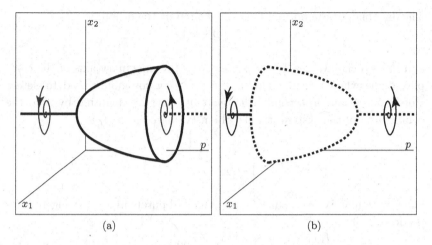

Fig. A.15 Hopf bifurcation: (a) supercritical; (b) subcritical.

unstable focus. When the parameter is varied the cycle shrinks until it collides with the equilibrium and after the collision only a stable equilibrium remains. By contrast, in the subcritical case the cycle is unstable and is the boundary of the basin of attraction of the stable equilibrium inside the cycle. Thus, after the collision there is only a repeller.

The normal form of the Hopf bifurcation is

$$\dot{x}_1(t) = px_1(t) - \omega x_2(t) + cx_1(t)\left(x_1^2(t) + x_2^2(t)\right),$$
$$\dot{x}_2(t) = \omega x_1(t) + px_2(t) + cx_2(t)\left(x_1^2(t) + x_2^2(t)\right),$$

which, in polar coordinates, becomes

$$\dot{\rho}(t) = p\rho(t) + c\rho^3(t),$$
$$\dot{\theta}(t) = \omega.$$

This last form shows that the trajectory spirals around the origin at constant angular velocity ω, while the distance from the origin varies in accordance with the first ODE, which is the normal form of the pitchfork. Thus, the stability of the cycle depends upon the sign of c, called *Lyapunov coefficient*.

Taking into account Figures A.13c and A.14c, it is easy to check that the Hopf bifurcation is supercritical [subcritical] if $c < 0$ [$c > 0$] (in the case $c = 0$ the system is linear and for $p = p^* = 0$ the origin is neutrally stable and surrounded by an infinity of cycles). For $p = p^*$ the origin of the state space is stable in the supercritical case and unstable in the opposite case.

The Jacobian of the normal form, evaluated at the origin, is

$$J = \begin{bmatrix} p & -\omega \\ \omega & p \end{bmatrix},$$

and its two eigenvalues $\lambda_{1,2} = p \pm i\omega$ cross the imaginary axis of the complex plane when $p = 0$. This is the property commonly used to detect Hopf bifurcations in second-order systems. In fact, denoting by $\bar{x}(p)$ the equilibrium of the system, the Jacobian evaluated at $\bar{x}(p)$ is

$$J = \begin{bmatrix} \dfrac{\partial f_1}{\partial x_1} & \dfrac{\partial f_1}{\partial x_2} \\[2mm] \dfrac{\partial f_2}{\partial x_1} & \dfrac{\partial f_2}{\partial x_2} \end{bmatrix}_{x=\bar{x}(p)},$$

and such a matrix has a pair of nontrivial and purely imaginary eigenvalues if and only if

$$\text{tr} J = \left.\frac{\partial f_1}{\partial x_1}\right|_{x=\bar{x}(p)} + \left.\frac{\partial f_2}{\partial x_2}\right|_{x=\bar{x}(p)} = 0,$$

$$\det J = \left.\frac{\partial f_1}{\partial x_1}\right|_{x=\bar{x}(p)} \left.\frac{\partial f_2}{\partial x_2}\right|_{x=\bar{x}(p)} - \left.\frac{\partial f_1}{\partial x_2}\right|_{x=\bar{x}(p)} \left.\frac{\partial f_2}{\partial x_1}\right|_{x=\bar{x}(p)} > 0,$$

where $\text{tr } J$ and $\det J$ are the trace and the determinant of matrix J.

In practice, for detecting Hopf bifurcations, one annihilates the trace of the Jacobian evaluated at the equilibrium and finds in this way the parameter values that are candidate Hopf bifurcations. Then, the test on the positivity of the determinant of J is used to select the true Hopf bifurcations among the candidates. Under suitable nondegeneracy conditions, the emerging cycle is unique and its frequency is $\omega = \sqrt{\det J}$, because $\sqrt{\det J} = \lambda_1 \lambda_2$, while its amplitude increases as $\sqrt{-c(p - p^*)}$.

Determining if a Hopf bifurcation is supercritical or subcritical is not easy. One can try to find out if the equilibrium is stable or unstable but this is quite difficult since linearization is unreliable at a bifurcation. Alternatively (but equivalently), one can determine the sign of the Lyapunov coefficient c transforming the system into the normal form via continuous change of variables and time scaling (see, *e.g.*, Guckenheimer and Holmes, 1997; Kuznetsov, 2004).

Tangent bifurcation of limit cycles

Other local bifurcations in second-order systems involve limit cycles and are similar to transcritical, saddle-node, and pitchfork bifurcations of equilibria. In fact, the collision of two limit cycles can be studied as the collision of the two corresponding equilibria of the Poincaré map defined on a Poincaré section cutting both cycles. Thus, the transcritical, saddle-node, and pitchfork bifurcations of such equilibria correspond to analogous bifurcations of the colliding limit cycles.

The most common case is the saddle-node bifurcation of limit cycles, more often called fold or tangent bifurcation of limit cycles, where two cycles collide for $p = p^*$ and then disappear, as shown in Figure A.16. On

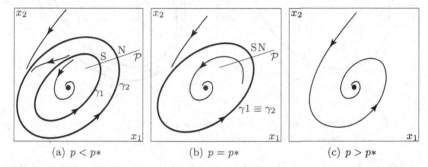

| (a) $p < p*$ | (b) $p = p*$ | (c) $p > p*$ |

Fig. A.16   Tangent bifurcation of limit cycles: two cycles γ_1 and γ_2 collide for $p = p^*$ and then disappear.

the Poincaré section $\mathcal{P}$ the bifurcation is revealed by the collision of two equilibria of the Poincaré map, S unstable and N stable, which then disappear. In terms of eigenvalue degeneracy, the eigenvalue of the linearized Poincaré map evaluated at S [N] is larger [smaller] than 1, so that when the two equilibria coincide, the eigenvalue of the unique linearized Poincaré map must be equal to 1.

Varying the parameter in the opposite direction, this bifurcation explains the birth of a pair of cycles, one of which is stable. While for Hopf bifurcations the emerging cycle is degenerate (it has zero amplitude), in this case the emerging cycles are not degenerate.

Flip (period-doubling) bifurcation

The flip bifurcation is the collision of two particular limit cycles, one tracing twice the other and therefore having double period, in a three- (or higher-) dimensional state space. In the supercritical [subcritical] case, it corresponds to a bifurcation of a stable [unstable] limit cycle of period T into a stable [unstable] limit cycle of period $2T$ and an unstable [stable] limit cycle of period T, as sketched in Figure A.17 (for the supercritical case) just before and after the bifurcation.

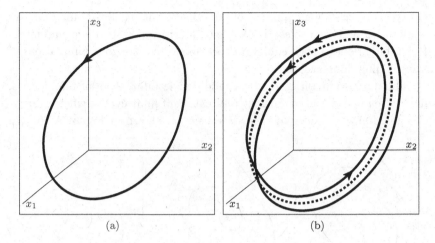

Fig. A.17   Flip bifurcation: (a) stable limit cycle of period T; (b) unstable limit cycle of period T and stable limit cycle of period $2T$.

Physically speaking, the stable limit cycle becomes only slightly different, but the key feature is that the period of the limit cycle doubles through the bifurcation. In other words, if before the bifurcation the graph of one of the state variables, say x_1, has a single peak in each period T, after the bifurcation the graph has two slightly different peaks in each period $2T$. On a Poincaré section, looking only at points of second return, the flip bifurcation resembles the pitchfork bifurcation, where two stable equilibria, $\bar{z}'$ and $\bar{z}''$ (corresponding to the two intersections of the period-$2T$ cycle with the Poincaré section), collide with a third unstable equilibrium $\bar{z}$ (the intersection of the period-T cycle) and disappear, while $\bar{z}$ becomes stable.

Mathematically speaking, the flip bifurcation is characterized by a multiplier of the period-T cycle equal to -1. In fact, when the cycle is unstable, the divergence from it, seen on a Poincaré section, is characterized by (first) return points which tend to alternate between points $\bar{z}'$ and $\bar{z}''$. This is due to a negative multiplier < -1. Just before the bifurcation, the cycle is stable but the multiplier is still negative, between -1 and 0, *i.e.*, the multiplier is equal to -1 at the bifurcation.

Neimark-Sacker (torus) bifurcation

This bifurcation, when supercritical, explains how a stable limit cycle can become a stable torus, by slightly varying a parameter. Figure A.18 clearly represents this bifurcation and shows that it can be interpreted (from right to left) as the collision of a stable vanishing torus with an unstable limit cycle inside the torus. On a Poincaré section, one would see a stable equilibrium (intersection of the cycle of Figure A.18a with the Poincaré section) bifurcating into an unstable equilibrium and a small regular closed curve (the intersection of the torus of Figure A.18b with the Poincaré section). In a sense, on the Poincaré section, one would observe invariant sets

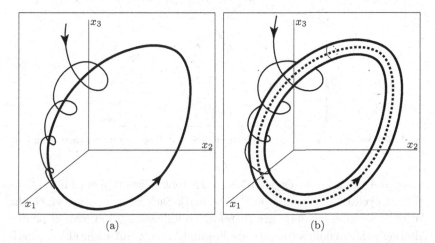

(a)   (b)

Fig. A.18   Neimark-Sacker bifurcation: (a) stable limit cycle; (b) unstable limit cycle and small stable torus.

with the same geometry as in the case of the supercritical Hopf bifurcation (see Figure A.15a). For this reason, the Neimark-Sacker bifurcation is sometimes confused with the Hopf bifurcation. Similarly, the subcritical Neimark-Sacker bifurcation resembles the subcritical Hopf bifurcation (see Figure A.15b).

In terms of cycle multipliers, the Neimark-Sacker bifurcation corresponds to a pair of complex conjugate multipliers crossing the unit circle in the complex plane. When the cycle is stable, nearby trajectories converge to the cycle by spiraling around it, while, when unstable, trajectories diverge from the cycle and spiral toward the torus.

In a two-parameter space, the Neimark-Sacker bifurcation curve separates the region in which the system has periodic regimes from that in which the asymptotic regime is quasi-periodic. However, as shown in Figure A.19, in the region where the attractor is a torus, there are very narrow subregions, each delimited by two curves merging on the Neimark-Sacker curve. In these subregions, called *Arnold's tongues*, the attractor is a cycle on

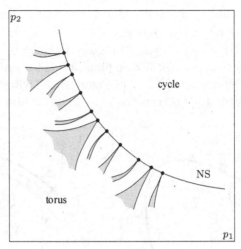

Fig. A.19   Neimark-Sacker bifurcation curve (NS) and Arnold's tongues emanating from it.

torus, and the two curves delimiting each tongue are tangent bifurcations of limit cycles. The points on the Neimark-Sacker curve from which the Arnold's tongues emanate are therefore codimension-2 bifurcation points. Although the Arnold's tongues are infinitely many, but countable (generically, there is a tongue for each possible $(r_1\!:\!r_2)$ pair characterizing a cycle on torus), only a few of them can be numerically or experimentally detected, as the others are too thin. Nevertheless, the Arnold's tongues are quite important because they explain the subtle and intriguing phenomenon known as *frequency locking*.

A.6   Global bifurcations

As already mentioned in Section A.4, global bifurcations cannot be detected through the analysis of the Jacobians associated with equilibria or cycles. However, they can still be viewed as structural changes of the characteristic frame.

Heteroclinic bifurcation

In Figure A.11 we have already shown the bifurcation corresponding to the collision of a stable manifold of a saddle with the unstable manifold of another saddle. This bifurcation is called heteroclinic bifurcation, since a trajectory connecting two saddles is called heteroclinic trajectory.

Homoclinic bifurcation

A special but important global bifurcation is the so-called *homoclinic bifurcation*, characterized by the presence of a trajectory connecting a saddle equilibrium with itself, called a homoclinic trajectory.

There are two collisions that give rise to a homoclinic trajectory. The first and most common collision is that between the stable and unstable manifolds of the same saddle, as depicted in Figure A.20. The second collision, shown in Figure A.21, is that between a node and a saddle whose unstable manifold is connected to the node. The corresponding bifurcations are called *homoclinic bifurcation to standard saddle*, or simply homoclinic bifurcation, and *homoclinic bifurcation to saddle-node*.

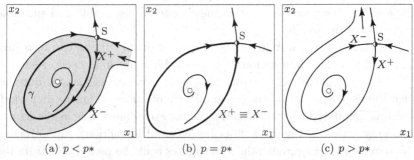

(a) $p < p*$ (b) $p = p*$ (c) $p > p*$

Fig. A.20 Homoclinic bifurcation to standard saddle: for $p = p^*$ the stable manifold X^- of the saddle S collides with the unstable manifold X^+ of the same saddle. The bifurcation can also be viewed as the collision of the cycle γ with the saddle S.

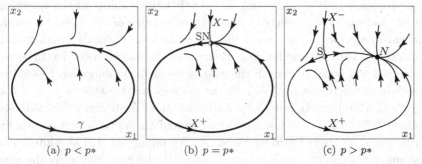

(a) $p < p*$ (b) $p = p*$ (c) $p > p*$

Fig. A.21 Homoclinic bifurcation to saddle-node: for $p = p^*$ the unstable manifold X^+ of the saddle-node SN comes back to SN transversally to the stable manifold X^-. The bifurcation can also be viewed as a saddle-node bifurcation on the cycle γ.

Figure A.20 shows that the homoclinic bifurcation to standard saddle can also be viewed as the collision of a cycle $\gamma(p)$ with a saddle $S(p)$.

When p approaches p^* the cycle $\gamma(p)$ gets closer and closer to the saddle $S(p)$, so that the period $T(p)$ of the cycle becomes longer and longer, since the state of the system moves very slowly when it is very close to the saddle. By contrast, Figure A.21 shows that the homoclinic bifurcation to saddle-node can be viewed as a saddle-node bifurcation on a cycle $\gamma(p)$, which therefore disappears. When p approaches p^* the system "feels" the forthcoming appearance of the two equilibria and therefore the motion of the state slows down close to the point where they are going to appear. Thus, in both cases, $T(p) \to \infty$ as $p \to p^*$ and this property is often used to detect homoclinic bifurcations through simulation. Another property used to detect homoclinic bifurcations to standard saddles is related to the form of the limit cycle which becomes "pinched" close to the bifurcation, the angle of the pinch being the angle between the stable and unstable manifolds of the saddle.

Looking at Figures A.20 and A.21 from right to left, we can recognize that the homoclinic bifurcation explains the birth of a limit cycle. As in the case of Hopf bifurcation, the emerging limit cycle is degenerate, but this time the degeneracy is not in the amplitude of the cycle but in its period, which is infinitely long. The emerging limit cycles are stable in the figures (the gray region in Figure A.20 is the basin of attraction), but reversing the arrows of all trajectories the same figures could be used to illustrate the cases of unstable emerging cycles. In other words, homoclinic bifurcations in second-order systems are generically associated with a cycle emerging from the homoclinic trajectory existing at $p = p^*$ by suitably perturbing the parameter. It is interesting to note that the stability of the emerging cycle can easily be predicted by looking at the sign of the so-called *saddle quantity* σ, which is the sum of the two eigenvalues of the Jacobian matrix associated with the saddle, *i.e.*, the trace of the Jacobian (notice that one eigenvalue is equal to zero in the case of homoclinic bifurcation to saddle-node). More precisely, if $\sigma < 0$ the cycle is stable, while if $\sigma > 0$ the cycle is unstable. As proved by Andronov and Leontovich (see Andronov *et al.*, 1973), this result holds under a series of assumptions that essentially rule out a number of critical cases. A very important and absolutely not simple extension of Andronov and Leontovich theory is Shil'nikov theorem (Shil'nikov, 1968) concerning homoclinic bifurcations in three-dimensional systems.

A.7   Catastrophes, hysteresis, and cusp

We can now present a simple but comprehensive treatment of a delicate problem, that of *catastrophic transitions* in dynamical systems. A lot has been said on this issue in the last decades and the so-called *catastrophe theory* (Thom, 1972) has often been invoked improperly, thus generating expectations that will never be satisfied. Reduced to its minimal terms, the problem of catastrophic transitions is the following: assuming that a system is functioning in one of its asymptotic regimes, is it possible that a microscopic variation of a parameter can trigger a transient toward a macroscopically different asymptotic regime? When this happens, we say that a catastrophic transition occurs.

To be more specific, assume that an instantaneous small perturbation from p to $p + \Delta p$ occurs at time $t = 0$ when the system is on one of its attractors, say $\mathcal{A}(p)$, or at a point $x(0)$ very close to $\mathcal{A}(p)$ in the basin of attraction $B\left(\mathcal{A}(p)\right)$. A first possibility is that p and $p + \Delta p$ are not separated by any bifurcation. This implies that the state portrait of the perturbed system $\dot{x} = f(x, p + \Delta p)$ can be obtained by slightly deforming the state portrait of the original system $\dot{x} = f(x, p)$. In particular, if Δp is small, by continuity, the attractors $\mathcal{A}(p)$ and $\mathcal{A}(p + \Delta p)$, as well as their basins of attraction $B\left(\mathcal{A}(p)\right)$ and $B\left(\mathcal{A}(p + \Delta p)\right)$, are almost coincident, so that $x(0) \in B\left(\mathcal{A}(p + \Delta p)\right)$. This means that after the perturbation a transition will occur from $\mathcal{A}(p)$ (or $x(0)$ close to $\mathcal{A}(p)$) to $\mathcal{A}(p + \Delta p)$. In conclusion, a microscopic variation of a parameter has generated a microscopic variation in system behavior.

The opposite possibility is that p and $p + \Delta p$ are separated by a bifurcation. In such a case it can happen that the small parameter variation triggers a transient, bringing the system toward a macroscopically different attractor. When this happens for all initial states $x(0)$ close to $\mathcal{A}(p)$, the bifurcation is called *catastrophic*. In contrast, if the catastrophic transition is not possible, the bifurcation is called *non-catastrophic*, while in all other cases the bifurcation is said to be *undetermined*.

We can now revisit all bifurcations we have discussed in the previous sections. Let us start with Figure A.13 and assume that p is small and negative, *i.e.*, $p = -\varepsilon$, that $x(0)$ is different from zero but very small, *i.e.*, close to the stable equilibrium, and that $\Delta p = 2\varepsilon$ so that, after the perturbation, $p = \varepsilon$. In case (a) (transcritical bifurcation) $x(t) \to \varepsilon$ if $x(0) > 0$ and $x(t) \to -\infty$ if $x(0) < 0$. Thus, this bifurcation is undetermined because it can, but does not always, give rise to a catastrophic transition. In a case like this, the noise acting on the system has a fundamental role since it determines the sign of $x(0)$, which is crucial for the behavior of the system after the parametric perturbation. We must notice, however, that in many cases the sign of $x(0)$ is *a priori* fixed. For example, if the system is positive because x represents the density of a population, then for physical reasons $x(0) > 0$ and the bifurcation is therefore non-catastrophic. However, under the same conditions, the transcritical bifurcation of Figure A.14a is catastrophic. Similarly, we can conclude that the saddle-node bifurcation of Figure A.13b is catastrophic, as also is that of Figure A.14b, and that the pitchfork bifurcation can be non-catastrophic (as in Figure A.13c) or catastrophic (as in Figure A.14c).

From Figure A.15 we can immediately conclude that the supercritical Hopf bifurcation is non-catastrophic, while the subcritical one is catastrophic. This is why the two Hopf bifurcations are sometimes called catastrophic and non-catastrophic. Finally, Figures A.16, A.20, and A.21 show that tangent and homoclinic bifurcations are catastrophic.

After a small parametric variation has triggered a catastrophic transition from an attractor $\mathcal{A}'$ to an attractor $\mathcal{A}''$ it is interesting to know if it is possible to drive the system back to the attractor $\mathcal{A}'$ by suitably varying the parameter. When this is possible, the catastrophe is called *reversible*. The most simple case of reversible catastrophes is the *hysteresis*, two examples of which (concerning first-order systems) are shown in Figure A.22. In case (a) the system has two saddle-node bifurcations, while

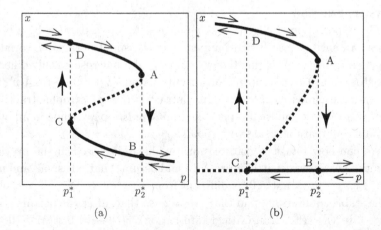

Fig. A.22 Two systems with hysteresis generated by two saddle-node bifurcations (a), and a saddle-node and a transcritical bifurcation (b).

in case (b) there is a transcritical bifurcation at p_1^* and a saddle-node bifurcation at p_2^*. All bifurcations are catastrophic (because the transitions $A \to B$ and $C \to D$ are macroscopic) and if p is varied back and forth between $p_{\min} < p_1^*$ and $p_{\max} > p_2^*$ through a sequence of small steps with long time intervals between successive steps, the state of the system follows closely the cycle $A \to B \to C \to D$ indicated in the figure and called the *hysteretic cycle* (or, briefly, hysteresis). The catastrophes are therefore reversible, but after a transition from $\mathcal{A}'$ to $\mathcal{A}''$ it is necessary to pass through a second catastrophe to come back to the attractor $\mathcal{A}'$. This simple type of hysteresis explains many phenomena not only in physics, chemistry, and electromechanics, but also in biology and social sciences.

An interesting variant of the hysteresis is the so-called *cusp*, described by the normal form

$$\dot{x} = p_1 + p_2 x - x^3,$$

which is still a first-order system, but with two parameters. For $p_1 = 0$ the equation degenerates into the pitchfork normal form, while for $p_2 > 0$ the equation points out a hysteresis with respect to p_1 with two saddle-nodes. The graph of the equilibria $\bar{x}(p_1, p_2)$ is reported in Figure A.23, which shows that for the parameters (p_1, p_2) belonging to the cusp region in parameter space, the system has three equilibria, two stable and one unstable (in the middle). In contrast with the hysteresis shown in Figure A.22, after a catastrophic transition from an attractor $\mathcal{A}'$ to an attractor $\mathcal{A}''$ (transition $B \to C$ in the figure), one can find the way to come back to $\mathcal{A}'$ without suffering a second catastrophic transition (path $C \to D \to A \to B$ in the figure).

A.8 Routes to chaos

The bifurcations we have seen in the previous sections deal with the most common transitions from stationary to cyclic regimes and from cyclic to quasi-periodic regimes. Only one of them, namely, the homoclinic bifurcation in third-order systems, can mark, under suitable conditions specified by Shil'nikov theorem, the transition from a cyclic regime to a chaotic one. In an abstract sense, the Shil'nikov bifurcation is responsible for one of the most known "routes to chaos", called *torus explosion*, characterized by the collision in a three-dimensional state space of a saddle cycle with a stable torus. Observed on a Poincaré section, the bifurcation is revealed by a gradual change in shape of the intersection of the torus with the Poincaré

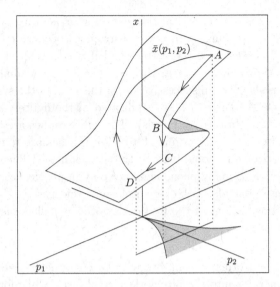

Fig. A.23 Equilibria of the cusp normal form. The unstable equilibria are on the gray part of the surface, which corresponds to the gray cusp region in the parameter space.

section, a shape which becomes more and more pinched while approaching the collision with the saddle cycle. After the collision, the torus breaks into a complex fractal set which, however, retains the geometry of a pinched closed curve, as shown in Figure A.24.

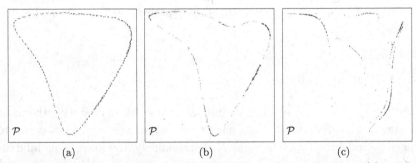

Fig. A.24 Torus explosion route to chaos viewed on a Poincaré section: (a) regular torus; (b) pinched torus; (c) strange attractor.

Another, perhaps better known, route to chaos is the *Feigenbaum cascade*, which is an infinite sequence $\{p_i\}$ of flip bifurcations where the p_i's accumulate at a critical value p_∞ after which the attractor is a genuine strange attractor. Very often, this route to chaos is depicted by plotting the local peaks of a state variable, say x_1, as a function of a parameter p, as shown in Figure A.25. Physically speaking, the attractor remains a cycle

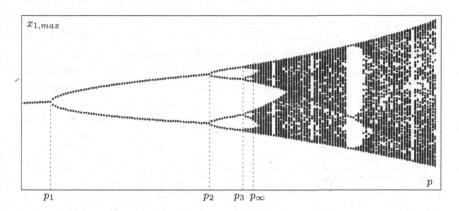

Fig. A.25   The Feigenbaum route to chaos: for $p \leq p_1$ the attractor is a cycle; for $p_1 < p < p_\infty$ the attractor is a longer and longer cycle; for $p \geq p_\infty$ the attractor is a strange attractor.

until $p = p_\infty$, but the period of the cycle doubles at each bifurcation p_i, while unstable (actually saddle) cycles of longer and longer periods accumulate in state space. This route to chaos points out a general property of strange attractors, namely, that they are basically composed by an aperiodic trajectory visiting a bounded region of the state space densely filled with saddle cycles, repelling in some directions (stretching) and attracting in others (folding).

A.9   Numerical methods and software packages

All effective software packages for numerical bifurcation analysis are based on *continuation* (see, *e.g.*, Beyn *et al.* (2002); Doedel *et al.* (1991a,b); Kuznetsov (2004), Chapter 10), which is a general method for producing in $\mathbf{R}^q$ a curve defined by $(q - 1)$ equations

$$F_1(w_1, w_2, \ldots, w_q) = 0,$$
$$F_2(w_1, w_2, \ldots, w_q) = 0,$$
$$\vdots$$
$$F_{q-1}(w_1, w_2, \ldots, w_q) = 0,$$

or, in compact form,

$$F(w) = 0, \quad w \in \mathbf{R}^q, \quad F : \mathbf{R}^q \to \mathbf{R}^{q-1}. \qquad (A.19)$$

Given a point $w^{(0)}$ that is approximately on the curve, *i.e.*, $F(w^{(0)}) \simeq 0$, the curve is produced by generating a sequence of points $w^{(i)}$, $i = 1, 2, \ldots$, that are approximately on the curve (*i.e.*, $F(w^{(i)}) \simeq 0$), as shown in Figure A.26a. The ith iteration step, from $w^{(i)}$ to $w^{(i+1)}$, is a so-called

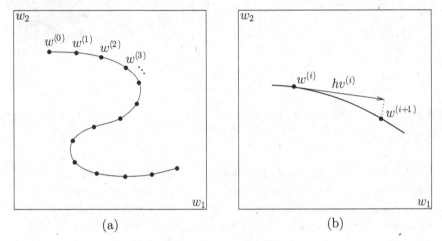

(a)   (b)

Fig. A.26   Generation of the curve defined by (A.19) through continuation.

prediction-correction procedure with adaptive step-size and is illustrated in Figure A.26b. The prediction $hv^{(i)}$ is taken along the direction tangent to the curve at $w^{(i)}$, where $v^{(i)}$ is computed as the vector of length 1 such that $\partial F/\partial w|_{w=w^{(i)}} v^{(i)} = 0$, the absolute value of h, called the step-size, is the prediction length, and the sign of h controls the direction of the continuation. Then, suitable corrections try to bring the predicted point back to the curve with the desired accuracy, thus determining $w^{(i+1)}$. If they fail, the step-size is reduced and the corrections are tried again until they succeed or the step-size goes below a minimum threshold at which the continuation halts with failure. By contrast, if corrections succeed at the first trial, the step-size is typically increased.

Given a second-order system $\dot{x} = f(x, p)$, where p is a single parameter, assume that an equilibrium $\bar{x}^{(0)}$ is known for $p = p^{(0)}$. Thus, starting from point $(\bar{x}^{(0)}, p^{(0)})$ in $\mathbf{R}^3$, the equilibria $\bar{x}(p)$ can easily be produced, as shown in Figure A.27, through continuation by considering (A.19) with

$$F(w) = f(x, p), \quad w = \begin{bmatrix} x \\ p \end{bmatrix}.$$

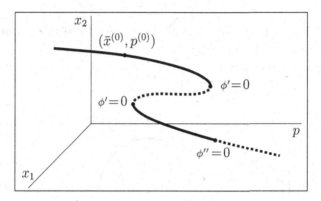

Fig. A.27   The curve $\bar{x}(p)$ produced from $(\bar{x}^{(0)}, p^{(0)})$ through continuation in the three-dimensional control space (p, x_1, x_2) and three bifurcation points, detected through the annihilation of the bifurcation functions ϕ' and ϕ''.

Moreover, at each step of the continuation, the Jacobian $J(\bar{x}(p), p)$ and its eigenvalues $\lambda_1(p)$ and $\lambda_2(p)$ are numerically estimated and a few indicators $\phi(\bar{x}(p), p)$, called *bifurcation functions*, are computed. These indicators annihilate at specific bifurcations, as shown in Figure A.27. For example, $\phi' = \det J$ is a bifurcation function of transcritical, saddle-node, and pitchfork bifurcations, since at these bifurcations one of the eigenvalues of the Jacobian matrix is zero and $\det J = \lambda_1 \lambda_2$. Similarly, $\phi'' = \operatorname{tr} J$ is a Hopf bifurcation function (see Section A.5). Once a parameter value annihilating a bifurcation function has been found, a few simple tests are performed to check if the bifurcation is really present or to detect which is the true bifurcation within a set of potential ones. For example, as clearly pointed out by Figure A.13b, at a saddle-node bifurcation the p-component of the vector tangent to the curve $\bar{x}(p)$ annihilates. By contrast, at transcritical and pitchfork bifurcations (see Figures A.13a and A.13c) two equilibrium curves, one of which is $\bar{x}(p)$, transversally cross each other, so that there are two tangent vectors at $p = p^*$, one with a vanishing p-component in the pitchfork case. Analogously, if $\phi''(\bar{x}(p^*), p^*) = 0$ one must first check that $\phi'(\bar{x}(p^*), p^*)$ is positive before concluding that $p = p^*$ is a Hopf bifurcation (see Section A.5).

Once a particular bifurcation has been detected through the annihilation of its bifurcation function ϕ, it can be continued by activating a second parameter. For this, (A.19) is written with

$$F(w) = \begin{bmatrix} f(x,p) \\ \phi(x,p) \end{bmatrix}, \quad w = \begin{bmatrix} x \\ p \end{bmatrix},$$

where w is now four-dimensional since p is a vector of two parameters. If the curve obtained through continuation in $\mathbf{R}^4$ is projected on to the two-dimensional parameter space, the desired bifurcation curve is obtained.

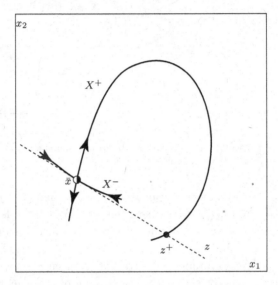

Fig. A.28   The bifurcation function $\phi = z^+ - \bar{x}$ is zero when there is a homoclinic bifurcation, *i.e.*, when the stable and unstable manifolds X^- and X^+ of the saddle collide.

In the case of local bifurcations of limit cycles and global bifurcations, the functions ϕ are quite complex and their evaluation requires the solution of the ODEs $\dot{x} = f(x,p)$. Actually, a rigorous treatment of the problem brings one naturally to the formulation of two-boundary-value problems (Beyn *et al.*, 2002; Doedel *et al.*, 1991b). For example, as shown in Figure A.28, homoclinic bifurcations can be detected by the function $\phi = z^+ - \bar{x}$, where z^+ is the intersection of the unstable manifold of the saddle $\bar{x}$ with the axis z passing through the saddle tangent to its stable manifold. Thus, ϕ is zero if and only if the saddle has a homoclinic connection.

There are many available software packages for bifurcation analysis, but the most interesting ones are AUTO (Doedel *et al.*, 2007), and MATCONT (Dhooge *et al.*, 2002). They can all be used to study systems with more than two state variables and they can detect and continue all bifurcations mentioned in this chapter. AUTO is the most popular software for bifurcation analysis and is particularly suited for the analysis of limit cycles, global bifurcations, and large scale systems. MATCONT, frequently updated, is more user-friendly and is developed in a MATLAB environment. It is less efficient but more effective than AUTO, as it can also detect and continue codimension-2 bifurcations.

Bibliography

Ahmad, W. M. and El-Khazali, R. (2007). Fractional-order dynamical models of love, *Chaos, Solitons & Fractals* **33**, 4, pp. 1367–1375.

Alligood, K. T., Sauer, T. D., and Yorke, J. A. (1996). *Chaos: An Introduction to Dynamical Systems* (Springer-Verlag, New York).

Andronov, A. A., Leontovich, E. A., Gordon, I. J., and Maier, A. G. (1973). *Theory of Bifurcations of Dynamical Systems on a Plane* (Israel Program for Scientific Translations, Jerusalem).

Austen, J. (1813). *Pride and Prejudice* (T. Egerton, Whitehall, London, England).

Bagarello, F. (2011). Damping in quantum love affairs, *Physica A* **390**, pp. 2803–2811.

Bagarello, F. (2012). *Quantum Dynamics for Classical Systems: With Applications of the Number Operator* (John Wiley & Sons).

Bagarello, F. and Oliveri, F. (2010). An operator-like description of love affairs, *SIAM Journal on Applied Mathematics* **70**, pp. 3235–3251.

Banerjee, M., Chakraborti, A., and Inoue, J.-i. (2015). Maximizing a psychological uplift in love dynamics, in R. Lopez-Ruiz, D. Fournier-Prunaret, Y. Nishio, and C. Gracio (eds.), *Nonlinear Maps and their Applications* (Springer), pp. 241–252.

Barley, K. and Cherif, A. (2011). Stochastic nonlinear dynamics of interpersonal and romantic relationships, *Applied Mathematics and Computation* **217**, pp. 6273–6281.

Baron, R. M., Amazeen, P. G., and Beek, P. J. (1994). Local and global dynamics of social relations, in R. Vallacher and A. Nowak (eds.), *Dynamical Systems in Social Psychology* (Academic Press, New York), pp. 111–138.

Bartholomew, K. and Horowitz, L. M. (1991). Attachment styles among young adults: a test of a four-category model, *Journal of Personality and Social Psychology* **61**, pp. 226–244.

Bauso, D., Dia, B. M., Djehiche, B., Tembine, H., and Tempone, R. (2014). Mean-field games for marriage, *PLoS ONE* **9**, 5, pp. e94933–1–15.

Bellomo, N. and Carbonaro, B. (2006). On the modelling of complex sociopsychological systems with some reasoning about Kate, Jules, and Jim, *Differential Equations and Nonlinear Mechanics* **2006**, pp. 1–26.

Bellomo, N. and Carbonaro, B. (2008). On the complexity of multiple interactions with additional reasoning about Kate, Jules, and Jim, *Mathematical and Computer Modelling* **47**, pp. 168–177.

Ben-Ze'ev, A. (2004). *Love Online: Emotions on the Internet* (Cambridge University Press).

Berne, E. (1970). *Sex in Human Loving* (Penguin Books Ltd).

Beyn, W.-J., Champneys, A. R., Doedel, E. J., Govaerts, W., Kuznetsov, Yu. A., and Sandstede, B. (2002). Numerical continuation, and computation of normal forms, in B. Fiedler (ed.), *Handbook of Dynamical Systems*, Vol. 2 (Elsevier Science, Burlington, MA), pp. 149–219.

Bielczyk, N., Bodnar, M., and Foryś, U. (2012). Delay can stabilize: Love affairs dynamics, *Applied Mathematics and Computation* **219**, 8, pp. 3923–3937.

Bielczyk, N., Foryś, U., and Płatkowski, T. (2013). Dynamical models of dyadic interactions with delay, *The Journal of Mathematical Sociology* **37**, 4, pp. 223–249.

Bonneuil, N. (2014). Emotions as dynamic systems in viability sets, *Mathematical and Computer Modelling of Dynamical Systems: Methods, Tools and Applications in Engineering and Related Sciences* DOI:10.1080/13873954.2014.961487.

Bowlby, J. (1969). *Attachment and loss: Vol. 1. Attachment* (Basic Books, New York), 2nd ed.

Breitenecker, F., Judex, F., Popper, N., Breitenecker, K., Mathe, A., and Mathe, A. (2008). Love emotions between Laura and Petrarch—an approach by mathematics and systems dynamics, *Journal of Computing and Information Technology* **4**, pp. 255–269.

Buder, E. H. (1991). A nonlinear dynamical model of social interaction, *Communication Research* **18**, pp. 174–198.

Buston, P. M. and Emlen, S. T. (2003). Cognitive processes underlying human mate choice: the relationship between self-perception and mate preference in western society, *Proceedings of the National Academy of Science* **100**, pp. 8805–8810.

Butler, E. A. (2011). Temporal interpersonal emotion systems: The "TIES" that form relationships, *Personality and Social Psychology Review* **15**, pp. 367–393.

Candaten, M. and Rinaldi, S. (2000). Peak-to-peak dynamics: a critical survey, *International Journal of Bifurcation and Chaos* **10**, 08, pp. 1805–1819.

Carbonaro, B. and Giordano, C. (2005). A second step towards a stochastic mathematical description of human feelings, *Mathematical and Computer Modelling* **41**, pp. 587–614.

Carbonaro, B. and Serra, N. (2002). Towards mathematical models in psychology: A stochastic description of human feelings, *Mathematical Models and Methods in Applied Sciences* **10**, pp. 1453–1490.

Carnelly, K. B. and Janoff-Bulman, R. (1992). Optimism about love relationships: general vs. specific lessons from one's personal experiences, *Journal of Social and Personal Relationships* **9**, pp. 5–20.

Colombo, A., Dercole, F., and Rinaldi, S. (2008). Remarks on metacommunity synchronization with application to prey-predator systems, *The American Naturalist* **171**, pp. 430–442.

Dercole, F. and Rinaldi, S. (2014). Love stories can be unpredictable: Jules et Jim in the vortex of life, *Chaos* **24**, pp. 023134-1–6.

Dhooge, A., Govaerts, W., and Kuznetsov, Yu. A. (2002). MATCONT: A MAT-LAB package for numerical bifurcation analysis of ODEs, *ACM Transactions on Mathematical Software* **29**, pp. 141–164.

Doedel, E. J., Champneys, A. R., Dercole, F., Fairgrieve, T. F., Kuznetsov, Yu. A., Oldeman, B., Paffenroth, R. C., Sandstede, B., Wang, X. J., and Zhang, C. H. (2007). AUTO-07p: Continuation and bifurcation software for ordinary differential equations, Department of Computer Science, Concordia University, Montreal, QC.

Doedel, E. J., Keller, H. B., and Kernévez, J.-P. (1991a). Numerical analysis and control of bifurcation problems (I): Bifurcation in finite dimensions, *International Journal of Bifurcations and Chaos* **1**, pp. 493–520.

Doedel, E. J., Keller, H. B., and Kernévez, J.-P. (1991b). Numerical analysis and control of bifurcation problems (II): Bifurcation in infinite dimensions, *International Journal of Bifurcations and Chaos* **1**, pp. 745–772.

du Toit, C. (2015). Polyphony and counterpoint: Mechanisms of seduction in the diaries of helen hessel and henri pierre roché, *Literator* **36**, 2, pp. 1–9.

Farina, L. and Rinaldi, S. (2000). *Positive Linear Systems: Theory and Applications* (John Wiley & Sons, Inc.).

Feichtinger, G., Jorgensen, S., and Novak, A. J. (1999). Petrarch's Canzoniere: Rational addiction and amorous cycles, *Journal of Mathematical Sociology* **23**, pp. 225–240.

Felmlee, D. H. and Greenberg, D. F. (1999). A dynamic systems model of dyadic interaction, *Journal of Mathematical Sociology* **23**, 3, pp. 155–180.

Ferrer, E. and Helm, J. L. (2013). Dynamical systems modeling of physiological coregulation in dyadic interactions, *International Journal of Psychophysiology* **88**, pp. 296–308.

Ferrer, E., Steele, J. S., and Hsieh, F. (2012). Analyzing the dynamics of affective dyadic interactions using patterns of intra-and interindividual variability, *Multivariate Behavioral Research* **47**, 1, pp. 136–171.

Fraley, R. C. and Shaver, P. R. (2000). Adult romantic attachment: Theoretical developments, emerging controversies, and and unanswered questions, *Review of General Psychology* **4**, pp. 132–154.

Frobenius, G. (1912). On matrices with nonnegative elements, *Sitzungsberichte der Deutsch Akademie der Wissenschaften, Berlin* **1**, pp. 456–477, in German.

Fruchter, G. E. (2014). Relationships in marketing and optimal control, in *Models and Methods in Economics and Management Science* (Springer), pp. 95–106.

Gardner, M. R. and Ashby, W. R. (1970). Connectance of large dynamic (cybernetic) systems: Critical values for stability, *Nature* **228**, p. 784.

Goldbeter, A. (1996). *Biochemical Oscillations and Cellular Rhythms: The molecular bases of periodic and chaotic behaviour* (Cambridge University Press).

Gonzaga, G. C., Campos, B., and Bradbury, T. (2007). Similarity, convergence, and relationship satisfaction in dating and married couples, *Journal of Personality and Social Psychology* **97**, pp. 34–48.

Gottman, J. M., Murray, J. D., Swanson, C. C., Tyson, R., and Swanson, K. R. (2002a). *The Mathematics of Marriage: Dynamic Nonlinear Models* (Bradford Book, Massachussetts Institute of Technology).

Gottman, J. M., Swanson, C. C., and Swanson, K. R. (2002b). A general systems theory of marriage: Nonlinear difference equations modeling of marital interaction, *Personality and Social Psychology Review* **6**, pp. 326–340.

Goudon, T. and Lafitte, P. (2015). The lovebirds problem: why solve hamilton-jacobi-bellman equations matters in love affairs, *Acta Applicandae Mathematicae* **136**, 1, pp. 147–165.

Gragnani, A., Rinaldi, S., and Feichtinger, G. (1997). Cyclic dynamics in romantic relationships, *International Journal of Bifurcation and Chaos* **7**, pp. 2611–2619.

Griffin, D. W. and Bartholomew, K. (1994). Model of the self and other. fundamental dimensions underlying measures of adult attachment, *Journal of Personality and Social Psychology* **67**, pp. 430–445.

Guckenheimer, J. and Holmes, P. (1997). *Nonlinear Oscillations, Dynamical Systems and Bifurcations of Vector Fields*, 5th edn. (Springer-Verlag, NY).

Hamermesh, D. S. (2011). *Beauty Pays* (Princeton Univeristy Press).

Hartl, R. F. and Mehlmann, A. (1984). Optimal seducing policies for dynamic continuous lovers under risk of being killed by a rival, *Cybernetics and Systems: An International Journal* **15**, pp. 119–126.

Hastings, A. and Gross, L. (eds.) (2012). *Encyclopedia of Theoretical Ecology* (University of California Press).

Hatfield, E., Brinton, C., and Cornelius, J. (1989). Passionate love and anxiety in young adolescents, *Motivation and Emotion* **13**, pp. 271–289.

Hatfield, E. and Sprecher, S. (1986a). Measuring passionate love in intimate relationships, *Journal of Adolescence* **9**, pp. 383–410.

Hatfield, E. and Sprecher, S. (1986b). *Mirror, Mirror: The importance of looks in everyday life* (State University of New York Press, New York).

Hazan, C. and Shaver, P. R. (1987). Romantic love conceptualized as an attachment process, *Journal of Personality and Social Psychology* **52**, pp. 511–524.

Hoppensteadt, F. (1974). Asymptotic stability in singular perturbation problems. II: Problems having matched asymptotic expansion solutions, *Journal of Differential Equations* **15**, pp. 510–521.

Huesmann, L. R. and Levinger, G. (1976). Incremental exchange theory: A formal model for progression in dyadic social interaction, in L. Berkowitz and E. Walster (eds.), *Advances in Experimental Social Psychology* (Academic Press, New York), pp. 191–229.

Johnson, D. (2010). *Love: Bondage or Liberation?: A Psycholological Exploration of the Meaning, Values and Dangers of Falling in Love* (Karnac Books).

Jones, F. J. (1995). *The Structure of Petrarch's Canzoniere: A Chronological, Psychological and Stylistic Analysis* (Brewer, Cambridge, UK).

Jørgensen, S. (1992). The dynamics of extramarital affairs, in G. Feichtinger (ed.), *Dynamic Economic Models and Optimal Control* (Elsevier Science), pp. 239–267.

Kirkpatrick, L. A. and Davis, K. E. (1994). Attachment style, gender, and relationship stability: A longitudinal analysis, *Journal of Personality and Social Psychology* **66**, pp. 502–512.

Koca, I. (2014). Mathematical modeling of nuclear family and stability analysis, *Applied Mathematical Sciences* **8**, 68, pp. 3385–3392.

Koca, I. and Ozalp, N. (2013). Analysis of a fractional-order couple model with acceleration in feelings, *The Scientific World Journal* **2013**, pp. 730736-1–6.

Koca, I. and Ozalp, N. (2014). On a fractional order nonlinear dynamic model of a triadic relationship, *Journal of Advances in Mathematics* **5**, 3, pp. 774–782.

Koss, L. (2015). Differential equations in literature, poetry and film, *Journal of Mathematics and the Arts* **9**, 1-2, pp. 1–16.

Kuznetsov, Yu. A. (2004). *Elements of Applied Bifurcation Theory*, 3rd edn. (Springer-Verlag, Berlin).

Levinger, G. (1980). Toward the analysis of close relationships, *Journal of Experimental Social Psychology* **16**, pp. 510–544.

Liao, X. and Ran, J. (2007). Hopf bifurcation in love dynamical models with nonlinear couples and time delays, *Chaos, Solitons & Fractals* **31**, 4, pp. 853–865.

Liu, W. and Chen, K. (2015). Chaotic behavior in a new fractional-order love triangle system with competition, *Journal of Applied Analysis and Computation* **5**, 1, pp. 103–113.

Lorenz, E. N. (1963). Deterministic nonperiodic flow, *Journal of the Atmospheric Sciences* **20**, 2, pp. 130–141.

Marsden, J. and McCracken, M. (1976). *Hopf Bifurcation and its Applications* (Springer-Verlag, New York).

May, R. M. (1972). Will a large complex system be stable? *Nature* **238**, pp. 413–414.

McDill, J. M. and Felsager, B. (1994). The lighter side of differential equations, *The College Mathematics Journal* **25**, pp. 448–452.

Mikulincer, M. and Shaver, P. R. (2003). The attachment behavioral system in adulthood: Activation, psychodynamics, and interpersonal processes, in M. Zanna (ed.), *Advances in Experimental Social Psychology* (Academic Press, New York), pp. 53–152.

Murray, J. D. (2002). *Mathematical Biology I: An Introduction*, 3rd edn. (Springer, New York, USA).

Novák, B. and Tyson, J. J. (2008). Design priciples of biochemical oscillators, *Nature Reviews Molecular Cell Biology* **9**, pp. 981–991.

Ozalp, N. and Koca, I. (2012). A fractional order nonlinear dynamical model of interpersonal relationships, *Advances in Difference Equations* **189**, pp. 510–544.

Pam, A. and Pearson, J. (1998). *Splitting Up: Enmeshment and Estrangement in the Process of Divorce* (Guilford Press).

Pikovsky, A., Rosenblum, M., and Kurths, J. (2001). *Synchronization: A Universal Concept in Nonlinear Sciences* (Cambridge University Press).

Radzicki, M. J. (1993). Dyadic processes, tempestuous relationships, and system dynamics, *System Dynamics Review* **9**, pp. 79–94.

Ramasubramanian, K. and Sriram, M. S. (2000). A comparative study of computation of Lyapunov spectra with different algorithms, *Physica D* **139**, pp. 72–86.

Rey, J.-M. (2010). A mathematical model of sentimental dynamics accounting for marital dissolution, *PLoS ONE* **5**, 3, pp. e9881–1–8.

Rey, J.-M. (2013). Sentimental equilibria with optimal control, *Mathematical and Computer Modelling* **57**, pp. 1965–1969.

Rinaldi, S. (1998a). Laura and Petrarch: An intriguing case of cyclical love dynamics, *SIAM Journal on Applied Mathematics* **58**, pp. 1205–1221.

Rinaldi, S. (1998b). Love dynamics: The case of linear couples, *Applied Mathematics and Computation* **95**, pp. 181–192.

Rinaldi, S., Della Rossa, F., and Dercole, F. (2010). Love and appeal in standard couples, *International Journal of Bifurcation and Chaos* **20**, pp. 2443–2451.

Rinaldi, S., Della Rossa, F., and Fasani, S. (2012). A conceptual model for the prediction of sexual intercourse in permanent couples, *Archives of Sexual Behavior* **41**, 6, pp. 1337–1343.

Rinaldi, S., Della Rossa, F., and Landi, P. (2013a). A mathematical model of "Gone with the Wind", *Physica A* **392**, pp. 3231–3239.

Rinaldi, S., Della Rossa, F., and Landi, P. (2014). A mathematical model of "Pride and Prejudice", *Nonlinear Dynamics in Psychology and Life Sciences* **18**, pp. 199–211.

Rinaldi, S. and Gragnani, A. (1998a). Love dynamics between secure individuals: A modelling approach, *Nonlinear Dynamics, Psycology and Life Sciences* **2**, pp. 283–301.

Rinaldi, S. and Gragnani, A. (1998b). Minimal models for dyadic processes: A review, *The Complex Matters of the Mind, ISBN* **1220298804**, pp. 87–104.

Rinaldi, S., Landi, P., and Della Rossa, F. (2013b). Small discoveries can have great consequences in love affairs: The case of Beauty and The Beast, *International Journal of Bifurcation and Chaos* **23**, pp. 1330038–1–8.

Rinaldi, S., Landi, P., and Della Rossa, F. (2015). Temporary bluffing can be rewarding in social systems: The case of romantic relationships, *Journal of Mathematical Sociology* **39**, pp. 203–220.

Roché, H.-P. (1953). *Jules et Jim* (Éditions Gallimard, Paris), (in French; English translation by P. Evans, Marion Boyars, 1998).

Rostand, E. (1897). *Cyrano de Bergerac* (Fasquelle, Paris, France), (in French).

Schachner, D. A. and Shaver, P. R. (2004). Attachment dimensions and sexual motives, *Personal Relationships* **11**, pp. 179–195.

Schuster, P. and Sigmund, K. (1981). Coyness, philandering and stable strategies, *Animal Behaviour* **29**, 1, pp. 186–192.

Shil'nikov, L. P. (1968). On the generation of periodic motion from trajectories doubly asymptotic to an equilibrium state of saddle type, *Mathematical USSR Sbornik* **6**, pp. 427–437.

Siljack, D. D. (1974). Connective stability of complex ecosystems, *Nature* **249**, p. 280.

Simpson, J. A., Lerma, M., and Gangestad, S. W. (1990). Perception of physical attractiveness: mechanisms involved in the maintenance of romantic relationships, *Journal of Personality and Social Psychology* **59**, 6, pp. 1192–1201.

Song, L., Xu, S., and Yang, J. (2010). Dynamical models of happiness with fractional order, *Communications in Nonlinear Science and Numerical Simulation* **15**, 3, pp. 616–628.

Sprecher, S. and Metts, S. (1986). Development of the 'romantic beliefs scale' and examination of the effects of gender and gender-role orientation, *Journal of Social and Personal Relationships* **6**, pp. 387–411.

Sprott, J. C. (2004). Dynamical models of love, *Nonlinear Dynamics, Psychology, and Life Sciences* **8**, pp. 303–314.

Sprott, J. C. (2005). Dynamical models of happiness, *Nonlinear Dynamics, Psychology, and Life Sciences* **9**, pp. 23–36.

Sternberg, R. J. (1986). A triangular theory of love, *Psychological Review* **93**, 2, pp. 119–135.

Strogatz, S. H. (1988). Love affairs and differential equations, *Mathematics Magazine* **61**, p. 35.

Strogatz, S. H. (1994). *Nonlinear Dynamics and Chaos* (Addison-Wesley, Reading, Massachussets).

Sunday, J., Zirra, D., and Mijinyawa, M. (2012). A computational approach to dynamical love model: The romeo and juliet scenario, *International Journal of Pure and Applied Sciences and Technology* **11**, 2, pp. 10–15.

Thom, R. (1972). *Structural Stability and Morphogenesis* (Benjamin, Reading, MA), (in French; English translation by Benjamin, 1975).

Tracy, J. L., Shaver, P. R., Albino, A. W., and Cooper, M. L. (2003). Attachment styles and adolescent sexuality, in P. Florsheim (ed.), *Adolescent Romance and Sexual Behavior: Theory, Research, and Practical Indications* (Lawrence Erlbaum Associates), pp. 137–159.

Vallacher, R. R., Read, S. J., and Nowak, A. (2002). The dynamical perspective in personality and social psychology, *Personality and Social Psychology Review* **6**, 4, pp. 264–273.

Wauer, J., Schwarzer, D., Cai, G. Q., and Lin, Y. K. (2007). Dynamical models of love with time-varying fluctuations, *Applied Mathematics and Computation* **188**, pp. 1535–1548.

Whyte, M. K. (1990). *Dating, Mating, and Marriage* (De Gruyter, New York).

Xu, Y., Gu, R., and Zhang, H. (2011). Effects of random noise in a dynamical model of love, *Chaos, Solitons & Fractals* **44**, 7, pp. 490–497.

Zhang, Z., Ye, W., Qian, Y., Zheng, Z., Huang, X., and Hu, G. (2012). Chaotic motifs in gene regulatory networks, *PLoS ONE* **7**, pp. e39355-1–11.

Zhao, Q. and Guan, J. (2013). Love dynamics between science and technology: some evidences in nanoscience and nanotechnology, *Scientometrics* **94**, 1, pp. 113–132.

Zhou, L., Shen, Y., Wu, W., Wang, Z., Hou, W., Zhu, S., and Wu, R. (2014). A model for computing genes governing marital dissolution through sentimental dynamics, *Journal of Theoretical Biology* **353**, pp. 24–33.

Zhuravlev, M., Golovacheva, I., and de Mauny, P. (2014). Mathematical modelling of love affairs between the characters of the pre-masochistic novel, in *2014 Second World Conference on Complex Systems,* (IEEE), pp. 396–401.

Index

Printed in the United States
By Bookmasters